Herbert Rees

Understanding
Product Design for
Injection Molding

Hanser Publishers, Munich Vienna New York

Hanser/Gardner Publications, Inc., Cincinnati

The Author:
Herbert Rees, 248386-5 Side Road (Mono), RR#5 Orangeville, Ontario, Canada L9W 2Z2

Distributed in the USA and in Canada by
Hanser/Gardner Publications, Inc.
6600 Clough Pike, Cincinnati, Ohio 45244-4090, USA
Fax: (513) 527-8950
Phone: (513) 527-8977 or 1-800-950-8977

Distributed in all other countries by
Carl Hanser Verlag
Postfach 86 04 20, 81631 München, Germany
Fax: +49 (89) 98 12 64

The use of general descriptive names, trademarks, etc., in this publication, even if the former are not especially identified, is not to be taken as a sign that such names, as understood by the Trade Marks and Merchandise Marks Act, may accordingly be used freely by anyone.

While the advice and information in this book are believed to be true and accurate at the date of going to press, neither the authors nor the editors nor the publisher can accept any legal responsibility for any errors or omissions that may be made. The publisher makes no warranty, express or implied, with respect to the material contained herein.

Library of Congress Cataloging-in-Publication Data
Rees, Herbert, 1915–
    Understanding product design for injection molding / Herbert Rees.
        p.   cm. – (Hanser understanding books)
    Includes index.
    ISBN 1-56990-210-0
    1. Injection molding of plastics.   I. Title.   II. Series.
TP1150.R453   1996
668.4' 12--dc20   96-32922

Die Deutsche Bibliothek - CIP - Einheitsaufnahme
Rees, Herbert:
Understanding product design for injection molding / Herbert
Rees. - Munich ; Vienna ; New York : Hanser : Cincinnati :
Hanser/Gardner, 1996
    (Hanser understanding books)
    ISBN 3-446-18815-0

© Carl Hanser Verlag, Munich Vienna New York, 1996
Typeset in the USA by M.D. Sanders, Herndon, VA
Printed and bound in Germany by Schoder Druck GmbH & Co. KG, Gersthofen

Rees

Understand

Hanser **Understanding** Books

A Series of Mini-Tutorials

Series Editor: E.H. Immergut

Introduction to the Series

In order to keep up in today's world of rapidly changing technology we need to open our eyes and ears and, most importantly, our minds to new scientific ideas and methods, new engineering approaches and manufacturing technologies and new product design and applications. As students graduate from college and either pursue academic polymer research or start their careers in the plastics industry, they are exposed to problems, materials, instruments and machines that are unfamiliar to them. Similarly, many working scientists and engineers who change jobs must quickly get up to speed in their new environment.

To satisfy the needs of these "newcomers" to various fields of polymer science and plastics engineering, we have invited a number of scientists and engineers, who are experts in their fields and also good communicators, to write short, introductory books which let the reader "understand" the topic rather than to overwhelm him/her with a mass of facts and data. We have encouraged our authors to write the kind of book that can be read profitably by a beginner, such as a new company employee or a student, but also by someone familiar with the subject, who will gain new insights and a new perspective.

Over the years this series of **Understanding** books will provide a library of mini-tutorials on a variety of fundamental as well as technical subjects. Each book will serve as a rapid entry point or "short course" to a particular subject and we sincerely hope that the readers will reap immediate benefits when applying this knowledge to their research or work-related problems.

E.H. Immergut
Series Editor

Preface

During the more than 40 years the author has worked designing molds for plastics products, he has had many occasions to view and discuss product designs with their designers. Usually, the contact with the designer was established only *after* a purchase order for a mold had been placed to produce a certain product. In reviewing the supplied drawings, and occasionally models, many questions had to be answered before starting to design and build a practical mold, suitable for the plastic, for the requirements of the product, and for the planned production and mold life expectancy.

Many product designers have excellent ideas when coming up with new products and improvements to existing ones, but often they apply design principles similar to those they were familiar with and had used with other materials, such as steel, glass, and others. In discussions, the product designers often have been shown to lack even some of the most basic understanding of the principles of plastics molding and of the behavior of plastics, both during molding and after the product is finished.

The present book, **Understanding** Product Design for Injection Molding, is intended to give the product designer some guidelines and examples on how to proceed once the idea for a new product has been conceived, or the decision has been made to redesign an existing product already made from plastic or from a different material.

It should be pointed out that any design must always be based on straight common sense; the designer cannot be shown *how* to design or to "invent" a product, but, once the general idea of a target has condensed sufficiently, the designer is advised on what to consider to make a viable plastics product.

There is much preliminary work to be done—research into existing (similar and/or related) products, market analysis, cost estimating, safety assessment, etc.—before the designer should start sketching his or her idea. But at this point, the designer should understand what can be done with plastics, which processes to select, how to choose the proper materials, and how to design so that the product not only will work as expected but also can be made efficiently.

It is hoped that this book will help the practicing product designer, and any student in this field, to acquire the necessary background required. The author has deliberately kept the language simple and, rather than fill the book with many charts, he has preferred to direct the reader to sources: the suppliers of plastics and other products required for the contemplated design. Many charts and other relevant documentation have the habit of becoming obsolete soon after they are issued, and it is better to get the latest data from the source.

Herbert Rees
Orangeville, Ontario

Contents

Introduction

This book is intended to acquaint the aspiring product designer with some of the basic rules and practices with which he or she should be familiar before attempting any product design.

Completely new designs are rare, although they may be necessary for truly new inventions (patented or not) which have no precedent, not even in similar fields. By definition, there can be no precedent for "new;" therefore, there can be more than one way to design and build a new product.

After deciding on a possible, practical, and acceptable target design, there are two paths open to the designer. The old way is to first experiment, with product and test prototypes, and then proceed with the necessary strength calculation and graphic delineation of the new design so that it can be produced—and perform—safely. The modern way, after a suitable design has been selected, is to use the preliminary drawings for stress and flow analyses; changes which then appear necessary are made before the final drawings are completed.

Today, most so-called "new" designs are really "improvements" of existing products or are based on similar, not necessarily related, products in other fields. (Note that patents are worded ". . . this invention describes an improvement of. . . .") It is common practice and highly recommended that, prior to any attempt to redesign or to simply copy an existing product, the designer should gather as many similar and related products as possible that are already on the market and study them carefully to determine which "good" features to retain and which "bad" features to avoid in the "new" product. This philosophy applies to any product, whether a simple handle or a complicated mechanism.

Most importantly, the designer must be aware of and consider possible alternatives when designing any product, and then answer all relevant questions before proceeding. The designer is responsible for ensuring that the product fulfills all expectations, both in performance and in appearance, *and that it can be produced economically.*

In the following chapters, the author highlighted many of the questions and decisions a designer will face while designing any product. Unfortunately, there are no rules for explaining or teaching "common sense"—the basic requirement for any design, whether for a consumer product or for a machine. The main principle is that the designer must understand where problems will and can arise, and that he or she will learn where to find the answers to such problems.

1 Plastic Product Design

Why design a product made of (any) plastic? Usually, the reason to design products made of plastic is either to create new products, never made before, or to create something similar to existing old products but which is better, more appealing to the user, or more economical to produce.

Typical examples of new products include compact disks (CDs) and enclosures for CDs, etc. There are many examples of old products which could be, or need to be, redesigned for the following reasons:

1. Different appearance (shape) and more sales appeal (decorations, colors). Typical examples include kitchenware, containers, cutlery (some previously made from plastics or other materials such as glass, wood, metal, paper, etc.).

2. Improved performance so that products last longer, are easier to use, and are easier to handle. For certain applications, some physical properties of plastics may be more suitable, such as corrosion and scratch resistance, proof against shattering, and lighter weight. Typical examples are eyeglass and contact lenses, and protective eye wear.

3. Weight (and cost) reduction of products previously made from plastics. Typically, weight reduction is most important in products used in the packaging industry and in the automotive and aircraft industries, where lower weights can result in significant fuel economy and allow greater payloads. Cost reduction, always important, is achieved by both reducing material costs and shortening molding cycles, while improving the product quality.

4. Changes in materials, for any of the above reasons. Typical examples are vials and bottles (from glass); hinges (from metal); floppy computer disk enclosures (from paper); hardware, such as knobs and tool handles (from wood); toys (from metal, paper, or cloth); automotive, aircraft, and naval parts (from cast zinc or aluminum, sheet steel, wood, or glass); and enclosures for electronic equipment (from wood or metal).

5. Reduction of the number of product components, usually achieved by combining functions. Typical examples include box and cover assemblies joined with "live hinges," the use of self-tapping screws to eliminate nuts, and staking of molded plastic bosses instead of metal rivets or screws.

6. Faster or easier assembly. A typical example is snap fit instead of screw assembly.

2   Making a New Product

How does a designer begin to make the contemplated new or redesigned product? First, the designer must decide which type of plastic to use for the contemplated product. Then, he or she must select the best method for processing the plastic to create the finished product.

2.1   Choosing the Proper Plastic

There are many thousands of plastics, each with specific chemical and physical characteristics which will affect the performance of the product. There can be vast differences in processing these plastics. First, the designer must choose between two basic types of plastic, *thermosets* and *thermoplastics*.

With thermosets, the raw material (resin) is "cured" under high pressure inside a *heated* mold. The chemistry of the resulting product is *different* from that of the original resin, and it cannot be recycled. A cured thermoset product (or any scrap) cannot be ground up to be reused, except perhaps as an *inert filler* for some other products. However, thermoset products inherently resist higher temperatures than most other plastics, and they have certain better physical, electrical, and chemical properties, as well.

Most thermosets are molded with the compression or transfer molding process (tires, dinnerware, electrical components, etc.). Some products, usually smaller ones, can be made using the injection molding process in specially adapted or modified injection molding machines. Some thermosets are used as liquids, applied to sheets or mats of glass and other fibers, and then cured under (relatively low) pressure and heat. (Examples are plywood, particle board, fiberglass structures, etc.)

With thermoplastics, the raw material must be heated before it is injected into a *cooled* mold. The resin does not change (or only minimally changes) its chemistry during the heating, molding, and cooling processes. Products (and any scrap) can be ground up and the material, in most cases, reused as if it were virgin material. Other processing methods are also possible, as discussed below.

A large variety of thermoplastics are molded using the injection molding process. Some products, such as disks for (analogue) music records, are molded in compression or in injection-compression molds, while (digital) compact disks are injection molded.

2.2   Selecting the Best Processing Method

Today, there are a number of different plastic processing methods, and new methods are developed frequently:

- Molding (compression, transfer, injection, injection-compression),
- extruding (structural shapes, tubing, sheet),
- extrusion blowing ("blow molding" or "bottle-blow molding"),
- injection blowing,
- stretch blowing (one- or two-step),
- expandable bead molding,
- reaction injection molding,
- structural foam molding,
- thermoforming,
- rotational molding, and
- lost-core molding (an area of injection molding).

Many of today's plastics may be suitable for only one or for just a few of the above-listed processing methods.

The type of product contemplated will play a large role in determining the method of manufacturing. The following discussion will describe some of the advantages and disadvantages of the various methods.

2.2.1   Molding

Processing under this category includes compression, transfer, or injection molding of thermosets or thermoplastics.

Advantages:

- Suitable for any size product, from those weighing fractions of a gram to large pallets and containers weighing 40 kg and more,

- excellent surface definition,
- good accuracy,
- good repetitiveness,
- high productivity, and
- product finishing usually is not required after molding.

Disadvantage:

- High mold and machine cost.

This book will be concerned mostly with designing products for injection molding, but, before continuing, we should consider some of the other plastic processing methods used.

2.2.2   Extruding

Extruding may only be performed with some thermoplastics. Typical products include endless tubing for food, industrial, agricultural, and medical use; garden hose; sewer and water pipe; structural profiles; sheet; and film. Extrusions can be rigid or flexible.

Advantages:

- Suitable for practically any size product, but limited by available equipment size,
- fair accuracy,
- good repetitiveness,
- high productivity, and
- low-cost tooling.

Disadvantage:

- Limited application as a final product.

2.2.3   Extrusion Blowing

In extrusion blowing, which is suitable only for some thermoplastics, cut lengths of hot extruded tubing, called *parisons*, are automatically transferred from an extruder to one or several cooled blow molds (blow cavities), where they are blown into the final shape.

Advantages:

- Suitability for all size bottles, toys, technical products (automotive fuel tanks, air ducts, etc.), and for containers, up to large drums and tanks,
- fair accuracy and repetitiveness,
- high productivity, and
- low-cost tooling.

Disadvantages:

- Limited quality of surface definition and
- product must be finished (deburred, etc.) after blowing.

2.2.4   Injection Blow Molding

Suitable for only some thermoplastics, injection blow molding involves hot, injection molded *preforms* that are transferred automatically into cooled blow molds where the preforms are blown into the final shape. This is only possible in special (injection blow) machines, or in special molds for standard injection molding machines.

Advantages:

- Suitability for very small to medium-size bottles and jars, drinking cups, etc.,
- good surface definition,
- high accuracy and repetitiveness, and
- high productivity.

Disadvantages:

- Special machine required, with relatively low mold cost, or
- high mold cost for use in standard injection molding machines.

2.2.5   Stretch Blow Molding

Most commonly used for molding PET bottles and jars, there are two methods of stretch blow molding: the 1-stage method and the 2-stage, or reheat and blow (R & B), method. The plastic is stretched to improve its physical properties.

2.2.5.1   1-Stage Method

Preforms are molded in a special machine and immediately, while still hot and without losing orientation, transferred first to a heat conditioning station and then to a blowing station where the preforms are stretched (optional) and blown. The main difference between this method and injection blow molding is that the preforms are heat-conditioned between the molding and blowing steps, in the same machine.

Advantages:

- Only one machine is required to do both molding and blowing,
- easier quality control, since each molding cavity corresponds to a specific blow cavity, and
- lower installation cost of total setup.

Disadvantages:

- Special machine required,
- number and spacing of injection cavities limited by the number and spacing of blow cavities possible for the size of machine,
- lower productivity, since the molding cycle controls the overall cycle (molding of the preform usually 3 to 6 times slower than the blowing cycle required to shape the bottle), and
- complicated machine with poor accessibility.

2.2.5.2   2-Stage or Reheat and Blow (R & B) Method

The preforms are molded in a standard injection molding machine. The *cold* preforms are then shipped randomly to a special stretch blow machine, where they are oriented, reheated, stretched, and blown.

Advantages:

- Standard injection molding machine can be used,
- molding cycles and temperature settings do not depend on blow speeds, etc., and can be optimized for highest productivity,
- molding plant need not be connected with blowing operation (Preforms can be shipped to the relatively simple reheat and blow equipment, located at a bottling plant, eliminating the costly shipping of bulky, lightweight bottles.),
- several highly productive (multicavity) molds can supply a few stretch blow machines, and

- possibility of stockpiling preforms, and blowing by drawing from stockpiles.

Disadvantages:

- Quality control is more difficult (The molding process must be perfect to ensure that only good preforms are received at the blowing machine. Due to the time delay between molding and blowing, any defects in preforms show up only a long time after molding.),
- slightly higher energy cost because preforms are first cooled completely, then reheated, and
- higher mold cost (However, due to the large quantities produced, the per unit cost is negligible.).

2.2.6   Expandable Bead Molding

Expandable bead molding is the molding of very lightweight, low density products for thermal insulation, energy-absorbing (consumer) packaging, drinking cups, meat trays and other disposable food containers, and products such as energy-absorbing automotive body parts, flotation devices, etc. This is an injection molding process that is limited to a few plastics. Very low pressures are used, and the molds are relatively simple and inexpensive.

2.2.7   Reaction Injection Molding

Also known as rigid foam molding, reaction injection molding traditionally included the molding of larger automotive and furniture components, but it also has been used lately for smaller products and strong, lightweight panels.

2.2.8   Structural Foam Molding

A blowing agent is injected with the plastic, to produce strong products with a solid outer skin and a foam interior.

2.2.9   Thermoforming

Only possible for some plastics, thermoforming includes the molding of large, rigid panels, such as refrigerator lining, but also small, lightweight disposable products such as drinking cups, food packaging, etc.

In this process, an extruded plastic sheet or strip is first heated, then placed into a forming station where it faces one or a number of cavities. Vacuum sucks the hot sheet into cooled cavities to give it the desired product shape. Occasionally, pressure plugs opposite the cavities assist the vacuum forming process. The sheet with the formed products is then moved to a trimming station where the products are cut from the sheet and removed for additional forming (rim beading, etc.) where required, and for stacking features. The web left over after trimming can be reprocessed. Thermoforming will not be further discussed.

2.2.10   Rotational Molding

Molding of large products such as storage tanks and containers, hollow automotive parts, complex toys, mannequins (display figures), etc., is achieved using rotational molding. Lately, there are more and more applications for this technology.

In rotational molding, plastic powder or liquid is placed inside a hollow mold, which is then closed, placed in an oven, and rotated biaxially while the plastic melts and adheres to the inside contour of the mold wall. After a period of time in a cooling chamber, the mold is opened and the product is removed. The molds are fairly simple, and there is no external (clamping) pressure involved; however, the cycles are long.

2.2.11   Lost-Core Molding

This molding process is used for products which require complicated cores *which cannot be removed* using any conventional injection molding technique. Typically, automotive products such as intake manifolds, which were previously sand cast from cast iron, can be made using this process.

The often quite complicated core is (injection) molded into one piece (or into several interlocking pieces) from a special metal alloy which has a very low melting point; this metal core (or core assembly) is then placed into an essentially conventional insert injection mold. After injection and cooling, the product, complete with the metal core, is put into a special oven where the metal is melted and removed from the finished product. The metal can then be reused.

3   Considerations for New Injection Molding Designs

There are many important points to consider when designing a new injection molded product. Before starting, consider the following requirements and parameters:

- Strength (weight/strength ratio),
- price (weight/price ratio),
- color, shape, and sales appeal,
- processibility,
- temperature properties,
- electrical and heat insulation properties,
- pollution and energy demands,
- life expectancy, and
- safety.

The product designer should be familiar with the selection and properties of available thermoplastics used for injection molding. This does not imply that the designer must know all the available plastics in this (very large) group, or be familiar with their physical and chemical characteristics, but he or she should know where to find the relevant information necessary for making a preliminary selection which might offer suitable properties for the product being considered.

After establishing the physical (and other) demands to which the product will be subjected, the designer can select, basing this choice on past experience and similar products, the plastic believed to be most suitable for the product under consideration. In case there is no such precedent, the designer will have to rely, as a first step, on the selection of a "promising" plastic by comparing physical and chemical properties shown in handbooks such as the *Encyclopedia* issued annually by *Modern Plastics*.

The designer also must be aware that physical and chemical properties alone are not sufficient for final selection of the plastic; it must be possible to process the plastic economically, as well. If it appears at this time that there is no plastic which suits the specified requirements and which can be injection molded, the designer is faced with three possibilities:

1. Approach the plastics material manufacturer to find out if there is a suitable plastic which for some reason is not shown in the listings, or if it is possible to compound such a plastic for the application on hand. Obviously, only if large volumes are considered will it be feasible and/or worth while for the resin supplier to create a "special" compound.
2. Abandon the use of injection molding, and select another processing method which will make it possible, using the selected plastic, to generate the product.
3. Abandon the use of plastic for the product, at this time.

Assuming one or more plastics are determined suitable for the demands of the contemplated product, the designer should approach one, or preferably several, manufacturers of the selected material to request their latest data sheets for any materials in the selected group. These data sheets will not only confirm (or revise) any data taken from the properties charts shown in handbooks or other data bases but also may state some examples in which such plastic was used. The data sheets also may offer some advice about product and mold design when using the plastic.In some cases, however, the information on moldability and mold design is not kept up to date and must be used with caution. For example, the manufacturer may describe results obtained using older types of molding equipment, and may not be aware of some more advanced molding techniques. Also, some molders (usually in cooperation with the machine builder and the materials supplier) may have devised their own methods of processing, which they may want to keep from becoming widely known.

Unless the new product is of a highly confidential nature, approach a molder who is well experienced in molding the type of product under consideration for advice regarding materials selection. This is an important and highly recommended step: eventually, the product design (product drawings) will have to be used to make molds, and unless the product designer is well versed in mold design, he or she should cooperate with an experienced molder and mold designer (in the appropriate field) before finalizing the product design and any detail drawings.

This cooperation can save much time (and frustration) by ensuring that the product is not only properly designed to accomplish what is expected of it but also that it can be molded, and further processed (assembled, decorated, etc.) after molding, without unexpected difficulties. In some cases, however, the designer may not want to cooperate with somebody else, and may wish to do all the experimentation in-house, regardless of the cost, to eventually take the advantage of being ahead of others by creating something entirely new. This can be a good policy if the investment in time and expenses for experimentation promises a substantial payback in the long run.

4 Designing a Product

In this context, "designing" means both original design and redesign. "Drawings" are produced on paper and printed as "blueprints" (white lines on blue background) or "white prints" (blue or grey lines on white paper), or as "hard copy," which are prints made on a "plotter" that reads directly from illustrations shown on the computer screen.

4.1 Graphic Delineation

This is the method used to express the design graphically (with drawings) so that others can understand and use them without additional *verbal* explanations. *Written* explanations (notes) on the drawings are usually acceptable, and often even desirable, if it is more practical to explain in a few words that which otherwise would require extensive or easily misunderstood graphic illustrations. But such notes must be short and concise, and written in an unequivocal language. Excessive use of notes can detract from the understanding of a design; the use of notes should be held to a minimum.

4.1.1 Drawing Board or Computer?

There is little difference in the method used to express a thought or concept for basic design, whether the old-fashioned drawing board or a modern computer-aided design (CAD) program is used. The designer's purpose is to describe the required shape of the product when committing it to paper or to the screen.

However, there is the question of speed and efficiency. When working with paper and pencil, using the most modern drafting machine, the actual drawing time is about the same as with a computer; however, it is faster to change, erase, and redraw lines with CAD.

4.1.1.1   Advantages of CAD

Simple but very time-consuming drafting chores, such as cross-hatching of sections, writing (labelling and numbering), and applying dimension lines and arrows, is much faster with CAD. If there are drawing and design standards in the memory of the CAD program, it will be much faster to generate basic design elements such as standard thread and hole sizes, fasteners, etc.; this can take much drudgery out of the design time.

It is also easier using CAD to generate additional views of the product, to add cross sections, and to highlight details, if required to better illustrate the product. A good product designer will always *keep the number of views and sections to a minimum*, but even so, several views may be required for a comprehensive description of the product.

For understanding the shape, the operation, and the functioning of any product, it is best to draw *"full scale"* (1:1); however, this is not always possible due to the size of the product. Therefore, one great advantage of CAD is that it very easily and quickly enlarges any areas of the product that are difficult to see or visualize, or that it can reduce the size of larger product drawings when required because of limited printer (plotter) and paper sizes available for making the final product drawings. The designer can easily work to full scale on any small portion of the overall design, or enlarge them if necessary, and then incorporate these detailed areas into a reduced, total illustration of the product.

CAD has the capability to electronically transmit the *product* design data to the system being used by the *mold* designer, provided the programs and computer systems used are compatible. The mold designer will save much time translating the product design into the mold design.

Another advantage of CAD is that the quality of delineation (line weight, shape and size of letters and numbers, etc.) is usually much better than that done by the hand of the average "draftsman."

"Designers" are often poor draftsmen and have difficulties in producing drawings that can be used by others (for checking, selling, manufacturing); their drawings must often be redrawn by "draftsmen" for better clarity. A good designer can usually express himself well verbally, and with simple hand sketches, but will often require somebody else to "put it on paper." With CAD, such sketches by the designer can be as easily, clearly, and cleanly produced on the screen as if they were made by a good draftsman.

4.2   Starting the Design Process

Before starting, the designer must have a complete list of questions about all the requirements which must be filled by the product. Typical questions are:

- Why do we need the new design?
- What must the product do (purpose or function)?
- What does the product replace?
- Are there competitive products? Are samples available?
- Are there any patents that would be infringed by a new design?
- How will the product be used?
- Where will the product be used (surroundings, climate, etc.)?
- Who will use the product (educational and/or cultural level of user)?
- Could the product be misused, creating danger or causing injury to the user or bystanders?
- Is the product disposable or "single-use" (throw away after one use)?
- Is the product seasonal?
- Is the product a promotional item (one time only)?
- Is the product expected to have a long time, frequent, repeated use?
- How many times will the product be reused (100, 1,000, 100,000, millions, etc.)?
- How important is product appearance (finish, colors, color match, etc.)?
- Is the product utilitarian or artistic?
- What is the expected product life (time to product obsolescence)?
- What postmolding operations are required (decorating, machining, etc.?
- What factors will affect product life (wear, corrosion, erosion, humidity, hot or cold temperatures, chemical action, electric arcing, etc.)?
- What special characteristics are required, such as electric resistance, optical clarity (transparency), etc.?
- How can the product be made (molding, blowing, etc.)?
- Can the product be made economically?
- What method of assembly, packaging, stacking, nesting, etc., will be used?
- Will the product require orientation (e.g., for assembly or painting)?
- How will the product be shipped? Boxes, transport containers, bulk?
- What is the expected cost range?

- When is it expected to be in production?
- Can the product be recycled? Does it require identification?

The more questions asked *and fully and correctly answered* before the start of the design work, the easier will be the actual designing process. Conscientiously using the answers to these questions will greatly reduce the design work.

It should also be made clear to the persons (the client or the boss) who can supply the answers that these questions are not asked out of *idle curiosity* but because the relevant answers can have an important, even decisive, influence on the design. There is nothing as embarrassing as finding out halfway through a design project that some of the given requirements or answers to these questions were wrong or incomplete, or that some key or even lesser requirements were overlooked or their importance was not fully appreciated.

After the need for a new (or revised) product has been justified and all requirements have been established by answering such questions as shown above, there are usually a *number of design possibilities* which can be immediately *excluded* for not fitting some of the above requirements. But, there are still a *large number of solutions possible which all could be acceptable.*

This is where the difference between a poor, a mediocre or a good designer becomes clearly evident. A poor or mediocre designer will, as a rule, contemplate the problem, then jot down the first possible or practical solution that comes to mind and continue to elaborate on it. This designer *may* be right in choosing the first solution, but, historically, this is seldom the case.

A *good* designer will contemplate several or sometimes a large number of *different possible design solutions* for any problem, and scrutinize each of them, to determine whether they all satisfy the answers to the questions asked. A good designer will also consider whether some portion of certain ideas in one layout or sketch could be joined with portions of some other sketches or ideas, to improve them, before settling on any one specific design.

It is also the mark of a good designer that he or she will *unhesitatingly* submit ideas, sketches, and drawings to peers, a supervisor, or others who might have something to contribute to the design and who could challenge the proposed designs. It is a common experience amongst designers that it is very difficult to spot one's own mistakes, while it is surprisingly easy for somebody else to pick out any errors in concepts or in the delineation, and/or to come forward with new ideas after somebody else's drawing has been viewed. It is a well-known fact that the *"second shot"* (by others) is always easier. Comments should never be considered a personal slight or a "nasty criticism."

The method of cooperative design using "concept" and "design" reviews is probably the best way to arrive at a good product. These concept and design reviews may at first appear cumbersome and time consuming. However, experience has proven again and again that this method is faster in the long run and yields better results (better designs) than any other method, because it does catch conceptual errors early in the project and takes advantage of the fact that several heads are usually better than one when it comes to creative ideas and their executions.

This does not mean that the original designer should do a hasty or sloppy job since ". . . it will all be taken apart by others anyway. . . ." After studying the underlying requirements, a good designer will initially come up a number of ideas and can then defend the rationale behind each idea presented and argue the validity of suggestions and criticisms by others. However, the designer must have an open and receptive mind and not be "defensive" while his or her designs are scrutinized or even condemned.

Presenting designers have often been disappointed or discouraged after hearing criticism from others, while fighting for their own ideas. In the long run, however, they will become much better designers who will have learned what matters in creating a good design.

In larger industries where a number of designers are available to do similar projects, it is common practice to alternate new projects between these designers and to have them all actively participate in design reviews, even if they are not the "lead" or presenting designer. Again, this will be to the benefit of all designers and, ultimately, to the project and the company.

4.3   Material Selection

When selecting plastics, the designer must understand that plastics often behave differently from nonplastic materials, particularly from metals. Plastics are often used to replace steel, other metals, glass, paper, etc. Frequently, one type of plastic will be replaced by another type which may be more suitable for the product, less expensive, or easier to convert (mold) into a product.

We will not discuss here the whole subject of characteristics of plastics but will alert the designer to differences in the behavior between plastics and other materials used previously for a similar product. If a similar product was previously made from a certain type of plastic, for instance, there is usually no problem in using the same or a related material again.

If, however, because of the user's or manufacturer's (molder's, assembler's) experience and/or because of new requirements, a different plastic may have to be used for the same application, *it may become necessary* to change the shape of the (existing) product. This could apply where the old product had insufficient mechanical strength or the requirements demanded of the older product have changed. The product may now need a similar but stronger plastic, or a different plastic altogether, which possibly cannot be produced in the identical (original) shape. For example, heavier wall thicknesses, more draft angles for molding, or some other modifications may be required to make the product more acceptable or "user friendly;" an altogether different method of manufacture (molded rather than blown, etc.) may even be required.

The above concerns also apply when *lightweighting* a product without significantly affecting its serviceability. Lightweighting may require thinning of the product walls or cutting large openings in the product walls to remove mass, provided neither the structural strength nor the product function will be affected by such openings. Often, some strategically placed ribs or other reinforcements may be required to make up for any loss in strength caused by the elimination of some excess plastic mass.

Lightweighting may also require changes in the processing characteristics of the plastic and the method of manufacture. Remember that, with molds, every reduction of the plastic mass means the *addition* of steel to the mold, which in most cases is *very difficult* to achieve. In such cases, usually either the cavity or the core, and frequently both, must be remade to match the new design. On the other hand, adding plastic is fairly easily done by recutting the mold steel (*removing* steel). In rare cases, adding steel by welding is possible and acceptable as a quick "fix," but this usually leaves visible marks on the surface of the finished product. Welding is not, in general, recommended for production molds, but it is frequently done in experimental molds where there is less emphasis on the appearance of the product.

4.3.1   Lightweighting

It must be understood that the cost of any product consists essentially of two major factors:

1. The cost of the plastic itself, and
2. the sum of all other costs, such as the mold cost attributed to the product, the machine hour cost, power, handling, overhead, etc.

In mass production of many plastic items, the first factor—the cost of the plastic—is often as high as 75% of the total cost of a product. For some products, such as disposable items, this percentage may be even higher. There is, therefore, a compelling incentive for the manufacturer to reduce the mass (weight) of any product, and more so when it comes to disposable items. As long as the product will meet the necessary demands, such as utility and safety, it should be acceptable in its lightest weight design.

Example 1 A typical example is the history of the disposable seven-ounce drinking cup, made from readily available and low-cost clear, crystal polystyrene (PS). From an original model of such a cup at 15 g per unit, the side walls, the rim, and the bottom were gradually reduced, while adding thin ribs not only for strength but to make it moldable at all with such a narrow thickness. This resulted in cups of about 9 to 11 g, which was better than before but still not good enough.

Further reduction of the mass of the cup was not possible, not only due to the limited capabilities of the injection unit (molding machine), which must provide extremely high pressure and very fast injection speed to fill the very thin walls before the plastic freezes, but also because the plastic used became so heavily flow oriented that the product became too brittle for the intended use. Instead of the readily available plastic, a better grade had to be developed, but, even then, a cup weighing 9 g was at the lowest practical limit.

However, when designers resorted to another method of manufacturing— in this case, injection blow molding— these obstacles disappeared. During the injection molding step, the rim and the portion *not to be blown* has an easily moldable thickness. The rest of the cup shape, which will be blown in the second step, is much thicker. This facilitates the first step by filling the thin rim through the heavier body walls. (Before, the heavy rim was filled through a very thin side wall, a condition which should generally be avoided in molding any plastic product.) After being transferred to the blow station, the heavy side walls and the bottom are blown to the final shape, resulting in extremely thin cross sections. While blowing, the plastic is biaxially oriented and thereby loses the brittleness formerly experienced. Even though it is now much thinner, the plastic is not brittle.

This example is given to highlight the necessary cooperation between the product and the mold designer, the materials supplier, the mold maker, the molding machine builder, and the molder to design a product which is more than a simple run-of-the-mill article. Typically, such cooperation applies whenever the quantities required make these efforts (high development costs) economically viable.

The result of the changes in the above-cited example was an excellent cup weighing only 6 g, or about 40% less than the original. Considering that one machine can produce 10,000 cups per hour, the redesign saved about 40,000 g, or 40 kg/hour. At a plastic cost of $1.00/kg, this is a substantial savings, considering that the machine runs at least 6,000 hours/year. The annual savings in material alone was $240,000/machine—more than enough to pay for the additional costs associated with designing and building special molds, machines, and suitable handling equipment.

Another example in the same product area shows that a certain product, such as a container, also can be produced using an entirely different method:

> *Example 2* Consider a margarine container molded from polypropylene (PP), which is easily injection molded. The product designer must consider several facts. First, the designer must be aware of the limits set by the injection molding machine in the area of injection capability. The thinner the product walls, the higher the injection pressures and injection speeds required to fill the product with a suitable resin that will remain strong enough for the purpose. Easier flowing materials will improve the molding capability, but there is a danger of losing required product strength. Higher resin cost is also a factor.
>
> A similar container could, however, be produced using an entirely different process—thermoforming—in which an extruded thin, wide strip of PP is reheated before entering the forming station and shaped in a multicavity forming mold into the desired shape. After forming, the strip advances to a trimming station; from there the products are conveyed for stacking, etc. The unused portion of the strip (the web) is then cut up and can be recycled.

This method is, in many cases, as good as injection molding, and is cost competitive. The designer must, however, be aware that there are certain disadvantages connected with this method of container production which concern the actual product:

1. The definition (the shape and sharpness of any engraved writing or decoration) of any injection molded product is much better than that created by thermoforming.

2. Injection molding permits better dimensional accuracy, which may be of important advantage for proper sealing (fit) of the matching lid.
3. Ribs to stiffen the rim are not possible with thermoforming, and any stacking features are usually less accurate than with injection molding.
4. For a long time, it was not possible or economical to thermoform crystal clear containers from plastics acceptable for the food industry; however, this drawback may change with new developments in resins.

If none of these points are important for the final shape of a new design, the designer has the choice to design either for injection molding or for thermoforming. The final decision should come from the molder, who will decide on economical grounds (availability of injection molding machines, investment in new machines or in a thermoforming machine line) which method to use. The actual product design drawing used will be slightly different for injection molding than for thermoforming, even though the main dimensions will be the same for either method.

Lightweighting has other advantages which are also very important:

·   The thinner the wall thickness, provided it is more or less uniform throughout the product, the faster can be the molding speed, and the higher the productivity.
·   There are additional savings in the cost of transporting the raw material, the finished goods, the cost of recycling, land fill costs, etc. All these costs represent savings in energy (electricity or oil) and result directly in a decrease in pollution.
·   The efficiency of an automobile or aircraft increases when their mass is reduced. The lighter the structure, the more efficient it is, and the greater are the possible payloads.

4.3.2   Mechanical Properties

When redesigning to replace a product previously made from other materials, such as steel, brass, aluminum, zinc or magnesium die cast metals, or wood, etc., the designer must understand that the characteristics of plastics in general cannot be replaced simply by comparing and extrapolating common benchmarks such as tensile strength.

4.3.2.1   Tensile Strength

Tensile strength of any material is the easiest characteristic to determine, since the process is to simply stretch a test sample until it breaks. Every material will go through certain stages before breaking.

When stretched slightly (up to the *elastic* limit) by the application of a load, the test piece will return to its original shape after the load has been removed. This can be shown on a stress/strain curve (Fig. 4.1). *Stress* is the force applied per unit cross section of the test piece (i.e., kg/cm^2 or psi). *Strain* is the increase in length per unit of length due to the stress applied (i.e., cm/cm or inch/inch).

For each *nonplastic* design material, the tensile strength is about the same over the normally expected temperature range of operation of the product. It is also, in general, not dependent on the *length of time* during which the product is stressed. Also, the tensile strength of materials such as steel is defined within a very narrow range of values, and is quite repetitive. For most (conventional) materials, the yield point, the elastic limit, and the proportional limit are very close to each other and are often used interchangeably.

For most thermoplastics, the tensile strength will vary greatly with changes in the temperature (see Fig. 4.2). Also, even for very similar plastics, the values for tensile strength can vary considerably, as can be seen in all charts and data sheets that show these values.

For plastics, the length of time the product is stressed is very important. If the plastic is stressed (within the elastic limit), *but only for a short time,* it will return to its original shape. If it is stressed for a *long time*, even well below the elastic limit, the molecular arrangement within the product will *gradually* change, or "creep," so that when the stress is eventually removed, the product will *not* return to its original shape.

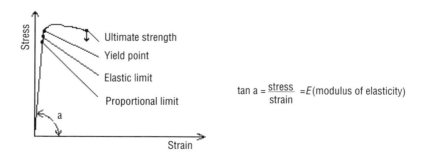

Figure 4.1 Typical stress/strain curve showing definitions of limits and modulus of elasticity

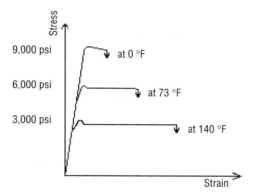

Figure 4.2 Schematic stress/strain curves illustrate the influence of temperature on the elastic limit of an injection molded plastic

An important effect of this behavior (creep) is that *a thermoplastic must not be used as a spring* if it requires even a small amount of preload, that is, if the plastic spring in its rest position remains stressed. However, it is quite practical to make use of the elasticity of any plastic if the spring portion of the product is not loaded after returning to its original (rest) position. If the spring stays any length of time (even a few days) in the fully loaded position, it will gradually relax (at the molecular level) and lose its elasticity and the desired spring effect.

A typical example where this springiness is used are the fuel tank filler caps of many motor cars. Plastic blades form a type of spring ratchet that will prevent the overtightening of the cap. The ratchets are relaxed, except while the cap is screwed onto the filler opening.

Another critical application where creep may have a significant negative effect is where screws are used in assembly of plastic parts. All screws depend on their elasticity to provide the necessary holding force. (For more on this subject, see *Mold Engineering* [1] by this author.) If a screw is tightened into a tapped hole, or even if self-tapping screws are used, the plastic is not strong enough to stretch the screw during torquing to provide the necessary preload to hold the assembly together against an expected separating force. The female (plastic) thread will be loaded under shear and bending. If the required holding force is minor, maybe just enough to hold a light cover in place, the plastic thread will usually be satisfactory. However, if the screw must be tightened strongly (I.e., to provide a good electrical contact between two metal surfaces held together by the screw), the plastic female thread will be under steady load and will gradually creep, eventually losing the force holding the two contact areas together (see also Section 5.2, Screw Assembly, p. 88).

The relationships showing the increase in strain over time (or loss of elasticity) for various plastics are available in *Engineering Properties of Thermoplastics* [2].

4.3.2.2   Compressive Strength

As with most metals and other nonplastics engineering materials, the compressive strength of plastics is about the same as the tensile strength.

4.3.2.3   Shear Strength

As with most other materials, the shear strength of most plastics is about 40–50% of the tensile strength.

4.3.2.4   Impact Strength

The impact strength of plastics varies over a wide range, depending on the type of plastic and on its molecular structure. Every material data sheet shows the impact strength values, together with the method that is used for the tests. Typical testing methods are Izod and Charpy tests (ASTM D-256), tensile impact tests (ASTM D-1822), drop tests, falling dart tests, etc. When comparing various plastics, it is important that the designer compare only those values which have been arrived at using the same methods of testing.

The same design rules apply as for metals or other engineering materials. Notches and holes, surface finish, any changes in wall thickness, etc., must be carefully considered to minimize the damage they can cause to the strength (life and performance) of the product. Some plastics are inherently more resistant to impact than others; some are very resistant, while others can be very brittle. *Fillers*, such as glass or carbon fibers, can add considerably to the impact strength.

The *temperature* during the operation of the product will have a much greater influence on the performance of the product than what is normally expected with other engineering materials. A plastic that may be suitable for an application at room temperature (22 °C or 70 °F) may be useless at a temperature of –18 °C (0 °F) or at a temperature higher than 60 °C (140 °F). Designers must be aware of this important fact when selecting a suitable plastic. (See also Fig. 4.2, p. 27.)

The effect of *humidity* also can be significant with certain plastics when the mechanical strength, and in particular the impact strength, depends on the amount of water absorbed (i.e., a typical example is nylon).

4.3.2.5   Flexural Strength

Flexural strength refers to the resistance of a plastic against bending. In fact, it is related to the tensile and compressive strength of the test specimen. The figures shown in charts are mostly useful in comparing the relative stiffness of materials.

4.3.2.6   Modulus of Elasticity (Tensile or Compressive)

The modulus of elasticity (E) is expressed as the slope (tan a) of the stress/strain curve up to the proportional limit (see Fig. 4.1, p. 27). The stiffer the material, the greater is the slope. For steel, on average, $E = 29,000,000$ psi, practically unchanged within a normal temperature range.

For example, for a strong plastic, such as unfilled polycarbonate molding compound, the values for E (tensile) = 345,000 psi, E (compressive) = 350,000 psi, and E (flexural) = 340,000 psi at room temperature decrease (or change) to 245,000 psi at 120 °C (250 °F). The designer must always refer back to current data sheets.

4.3.2.7   Hardness

Hardness, scratch and abrasion resistance, and wear (friction) properties are characteristics which can be important for the selection of the most suitable material for a specific job. For the purpose of comparison, the designer must make sure that the data shown in the various charts have been produced using the same test methods.

In summary, regarding mechanical properties, it is important that the designer scrutinize the material data sheets for all peculiarities of a plastic before settling on a final specification; additional information on critical properties (some published, some not) may be available from materials suppliers. Some of these suppliers may have newly developed data, and it is up to the designer to find out from different sources what is available for the application. The suppliers are always very interested in finding new markets for their products and are usually of great help.

It does not make much sense to include property tables or charts of all plastics in this book. There is ongoing development of completely new and modified resins, and engineering information which was accurate last year may be superseded now. Usually, the best sources are up-to-date data sheets issued by the materials suppliers.

However, charts shown in annual publications such as the *Encyclopedia of Plastics* [3] are a good start for directing the designer who has no idea which plastic to select where there is no precedent. When contacting material specialists from a few (competing) suppliers, the designer can request the latest data sheets and may also obtain additional information about newer, potentially more suitable, materials.

The designer must always keep an open mind. The fact that a material has been used a long time for a certain application does not necessarily mean it is the best choice, even if it was acceptable in the past. There now may be newer, better materials available, maybe at lower cost, or with easier processing characteristics or other advantages in the areas of colors, transparency, wear, etc.

Unless there is a compelling economical, sales, or engineering reason, it is usually unnecessary to revise the material for an existing product. But a new product planned to replace an older one (or to compete with it) should certainly be considered a candidate for any new material available.

4.3.3   Physical Properties

The same considerations listed above under mechanical properties apply here. Typical physical properties are:

- Coefficient of thermal expansion,
- deflection temperature under flexural load,
- thermal conductivity,
- electrical properties (conductivity, dielectric strength, arc resistance),
- resistance to flame,
- resistance to UV (ultra violet) light,
- resistance to water (cold, hot, steam),
- resistance to solvents, such as hydrocarbons or other organic solvents, and
- resistance to inorganic agents and solutions, such as weak acids, strong acids, weak alkalies, strong alkalies, or salts.

All plastics properties charts show how each plastic is affected when subjected to the influence of the above listed or other conditions and/or agents. If the product to be designed is subject to conditions not listed in the charts, the designer must obtain the information from the supplier of a material that may have been selected for the job. Also, the designer may want to have the supplier *quantify* such expressions as "weak" acids: How weak? In what solution or percentage?

4.3.4   Processing Methods

The importance of knowing how the product will be made was stressed earlier in Chapter 2. Typical methods today include:

- Molding,
- blowing,
- forming,
- casting, and
- machining from the solid, among others.

The product design for each of these methods will be different, to suit each method.

4.4   Product Shape

Every product has its own characteristic shape. The "shape" may be a simple, flat strip or disk, or a complicated, three-dimensional shape incorporating many features which are desired in the product. Certain "rules" must be kept in mind by the product designer with respect to product shape:

1. The shape must accomplish all that is expected of the product and provide adequate strength for its use.

    This is a very important stipulation that is often overlooked the first time around and which often, unfortunately, requires redesign shortly after the product is "out in the field" (i.e., after being seriously tested by the end users). Only too often, strength is thought to be increased by simply indiscriminately adding more plastic to the thickness. However, this can result in a serious waste of plastic. As discussed earlier, the cost of plastic can represent a very large portion of the cost of any molded product; also, the thicker the plastic, the slower is the molding cycle (and the production), thus adding to the cost of molding (and the product).

2. The shape must be so designed that it can be produced readily (and economically) by the method intended for its manufacture.

    For example, for various design reasons, it may be desirable to make a hollow product by *injection* molding. However, for certain plastics and products, this may not be possible at all, or else certain complicated, slow, and costly methods would be required: the hollow form could be assembled, by using more than one molded piece joined together; collapsible cores might be used, resulting in a delicate and costly mold; or possibly lost-core molding, a very expensive molding technology, might be employed.

3. The shape must be designed for appearance so that the product has sales appeal and is pleasing to the end user, where applicable. From past design experience, it can be said that ". . . a good-looking shape is also a well-functioning shape. . . ."

4. The product shape should include as many features and functions as possible to make the product attractive for use, easy (and at low cost) to mold, and to reduce the costs of any further steps in the manufacturing process where the molded product will be used.

Typical features appearing in the molded product are:

·   Openings (inlets, outlets, passages for electrical devices, openings in walls to reduce total mass, where possible, ventilation slots and "louvers");

·   bosses;

·   provisions for methods of assembling other components or hardware;

·   provisions for assembling other plastic components (covers, etc.) by snap fit, or by sliding into place;

·   handles and grips for easy handling of the product by the end user, stops to restrict motions, etc.;

·   studs for mounting other parts;

·   registers;

·   dials or escutcheons;

·   lettering and logos;

·   "knock-outs" (provisions in the product wall for future openings, to be broken through by user);

- springs;
- dowels and other locating devices,

and many more.

In the remainder of this section, we will discuss some principles and guidelines to consider when designing the shape of a product, which usually consists of a solid, homogeneous outer layer of molded plastic, often referred to as the "wall." Note that new technologies such as co-injection allow the wall to consist of more than one layer of plastic to enable the use of some low-cost, possibly reground, plastics together with higher cost virgin, or even completely different materials with different physical and other properties. There are also occasions where, using insert molding techniques, the outer wall is entirely or in part some other material, such as steel, printed cardboard, etc., and the plastic is used just to create some or all of the joints or to add some of the features usually associated with all-plastic products. Typical features such as openings, louvers, holes, reinforcements such as ribs and gussets, provisions for mounting such as studs and hubs, etc., will be discussed later.

4.4.1   Outside Surface

There is always a surface which defines the shape the product. This surface can be:

1. A flat strip of any shape, with or without holes, etc. (Fig. 4.3, left).
2. A shallow box (e.g., a container or a lid, Fig. 4.3, right) for packaging, medical, or technical products. The outside surface is the bottom and the side walls of the box.

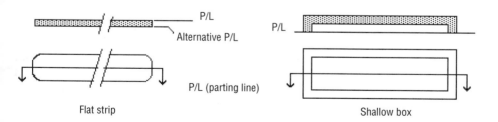

Figure 4.3 Schematic drawings showing product shape and parting lines for a flat strip (left) and a shallow box (right)

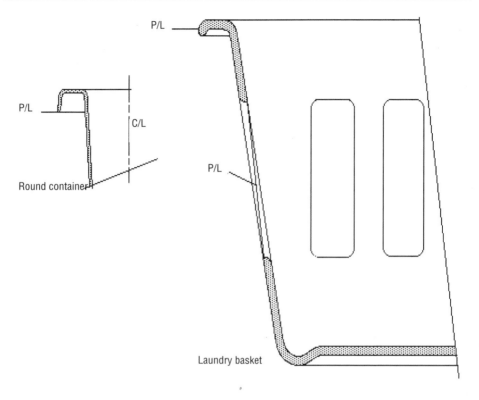

Figure 4.4 Schematic drawings showing product shape for a round container, such as a pail with a flange (left) or a laundry basket with a flange and openings (right)

3. A deep container (e.g., a large pail with a flange) or a laundry basket (Fig. 4.4), etc. The outside surface consists of the bottom, the sides wall, and any flange extending at the rim.

4. Any enclosure, such as a TV cabinet, housing for electric hand tools, etc. The outside layer (or wall) of the molded product may be flat or curved; in general, it may have any shape suitable for the job.

Having stated the above, the designer must also be aware that, today, with the frequent use of numerically controlled (NC) machine tools for metal working, it is of advantage for manufacturing if the shape selected can be expressed in geometric terms, which can be generated in the program supplied to these NC machines. The old-fashioned method of working from models, using duplicating (tracing) mills or lathes, or pantographs, is rarely used today, except where artwork is transformed into cavity work, as in the toy industry.

It also is important for the designer to be aware that a really flat surface usually is very difficult to achieve in molding, especially with high shrinkage plastics, mainly due to the difficulty of equally cooling both sides of such a surface. Wherever possible, a slight curvature is always preferable, with enough tolerance to ensure that any variations of the curvature will fall within the specified limits. If a really flat surface is required, it may need much longer molding cycles.

4.4.1.1   Plastic Thickness

Rule 1: A molded piece should have uniform thickness throughout.

Uniform thickness is possible to achieve but is rarely the case; in fact, in most cases, it is impossible. However, the designer should make every effort to make the thickness in any area of any product as uniform as possible. The reasons are quite simple (refer to *Mold Engineering* [1] by this author):

- The thicker the plastic, the slower it will cycle in the machine. The thickest portion controls the cycle time, since the product cannot be ejected from the molding machine before it is reasonably cool and stiff. As a rule, the increase in cycle time is not just linear with the thickness, but is greater. In other words, twice the wall thickness will require more than twice the cooling time. (How much more time depends greatly on the mold design.)
- The shrinkage of a heavier section is greater than that of a thinner section. A heavy section will have *voids* (empty spaces inside) and *sinks* (molding flaws outside). This condition worsens as the shrinkage factor of the plastic increases; in other words, a plastic with high shrinkage, such as PP, PE, or nylon, will be worse off with heavy sections than a plastic with lower shrinkage. (Note that the pressure inside the void is negative, or a vacuum.)

Figure 4.5 (next page) illustrates the effect and the meaning of a void and a sink. Note that while sinks and voids occur mostly in plastics with high shrinkage, voids can also be present in thick sections of moldings made from plastic with low shrinkage.

- The effect of varying shrinkage within one shape can result in slower molding cycle times, warp, and loss of dimensional accuracy and stability of the product after molding. It is possible to mold such

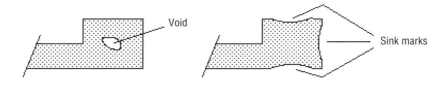

Figure 4.5 Shrinkage during production may occur as a void (left) or as sink marks (right)

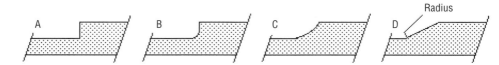

Figure 4.6 Various transitions between thicknesses in a product: A. "notch" effect, a severe stress riser; B. a small radius at the corner, C. a large radius at the corner, and D. a slope with a large, unnotched radius

heavy sections if they are necessary, but it will require special care in the selection of melt temperatures, cooling temperatures, and injection and hold pressures, and will usually result in slower molding cycles.

- Every transition from thinner to heavier thickness creates stress risers and an inherent weakness in the plastic product. This is especially serious if the transition is radical, such as a step.

In Fig. 4.6A, the product shows a "notch effect," a severe stress riser causing a serious weakness in the corner. A small radius (Fig. 4.6B) is better, and a large radius (Fig. 4.6C) or a slope (Fig. 4.6D) is best for strength in the corner area, provided the slope blends into the straight portion with a large radius and without a notch.

If, for strength reasons, the thickness must be increased, possibly after calculating the stresses on the product under the anticipated load, the designer should try to stay with a thickening of less than 25% and not, as shown in the above sketches (Figs. 4.5 and 4.6) a thickening of 100%. This 25% is an value based on experience and is about the limit up to which the molding characteristics will not change noticeably (i.e., the cooling time may not be affected by such thickening, and the notch effect will be small, provided the transition is gradual, over as large a distance as possible, with a long slope).

4.4.1.2 Gate Location

Rule 2: Make sure that the mold can (and will) be gated so that the thickness of the plastic from the gate toward the remote areas either remains constant or diminishes.

To meet this requirement, the product designer may want to seek the assistance of an experienced mold designer.

Not all products adhere to this rule, but the product designer must understand the flow of plastic within the mold. The plastic pressure is highest near the gate, and, due to the restriction in the space between cavity and core, the injection pressure drops as the cavity space is filled with plastic farther away from the gate. This means that, by the time the plastic reaches an area where more plastic is required (such as in a thickening of the product), the pressure available to fill that area is low, and it will be more difficult to fill the cavity and to pack out the product to specification.

(Note that for all plastics, areas filled under low pressure will shrink more than those filled under high pressure, regardless of the shrinkage values. For example, a PP container may shrink 1.5 % near the gate in the center of the base, but it may shrink closer to 2.0 % near the rim, remote from the gate.)

Of course, such more difficult products can be, and are being, molded satisfactorily, but they require higher injection pressures and higher melt temperatures, which may require stronger molds and heaver (larger) molding machines, which definitely will require more cooling time to bring the higher plastic temperatures down to where the product can be ejected. (Also, more energy is required to provide the additional heat for the melt and then to remove this additional heat during the cooling process.) In addition, the molding cycle will have to be extended accordingly.

Although there are often reasons why it appears that large changes in thickness are unavoidable, the designer should still consider whether they can somehow be avoided. Consider the following examples (Fig.4.7):

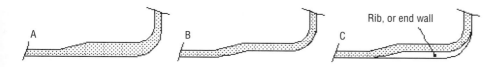

Figure 4.7 Extreme thickness changes in A should be avoided; thickness is almost constant in B but the product shape is changed; the thickness in C is uniform and the added rib keeps the product shape unchanged

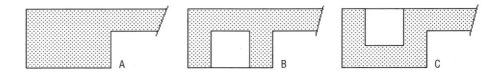

Figure 4.8 Excessive thickness in a product (A) can be eliminated by coring out from the cavity (B) or the core (C) side of the product

The shape shown in Fig. 4.7A may be required for the purpose of the product, but it is obviously not a shape most suitable for molding, nor for equal strength. Instead, the designer can choose between Figs. 4.7B and 4.7C, according to the needs of the product. In Fig. 4.7B, the thickness remains more or less constant, staying well within the 25% "permitted" before affecting the moldability and product strength. The silhouette of the product, however, is changed. In Fig. 4.7C, the silhouette of the product is maintained, either by addling flanges at the ends of the step or by adding ribs of the shape shown.

Similarly, excessive thickness can often be avoided by coring the thick section from one side or from the other, whichever is more suitable for the product (and its appearance). In contemplating such coring out of any heavy section, it must be a basic consideration for the product designer to avoid a waste of plastic, to reduce the molding cycles, and to avoid excessive, uneven shrinkage, as well as sinks and voids, to save power and to increase productivity. Figure 4.8 shows two typical examples of coring.

Obviously, Fig. 4.8A is bad design, since the change in thickness is far more than 100%. In Fig. 4.8B and 4.8C, however, either design is good, and the choice depends entirely on which better suits the desired product and, to some extent, on the appearance. In most cases, coring from the core side of the mold (the inside of the product) is preferable. The effect of coring will also help to keep the product on that side of the mold from which it will be ejected—usually, but not necessarily, the core side of the product.

4.4.1.3   Flow Path

Rule 3: Plastic flow in a mold always takes the path of least resistance.

The following two sketches (Fig. 4.9) illustrate how important it is for the product designer to understand the flow of plastic within the cavity space.

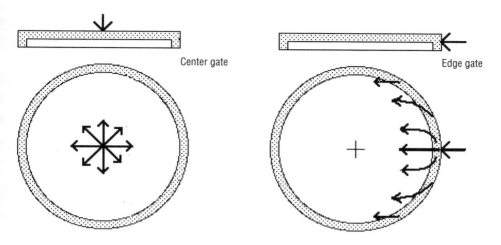

Figure 4.9 Schematic drawings of a shallow, round dish show plastic flow with center gating (left) and with edge gating (right)

(Obviously, there is an infinite number of design possibilities where these problems can occur.)

A shallow, round dish is shown in Fig. 4.9 with a uniform thickness at the base and at the rim. This is an ideal condition and is often possible in round containers, lids, etc.

With center gating (Fig. 4.9 left), the plastic will flow uniformly from the center toward the rim. This is a preferred gating method, but it is not always acceptable; for example, if the flat bottom must be free of any gate vestige, center gating cannot be used.

With edge gating (Fig. 4.9 right), the plastic will flow approximately as illustrated by the arrows and will eventually converge at the edge opposite the gate. There is no problem with venting.

In Fig. 4.10, the rim is thicker than the bottom. In a good design, this should, but often cannot, be avoided. The major two resultant problems are illustrated.

With center gating (Fig. 4.10 left), the filling flow is similar to that of the first example (Fig. 4.9 left), and the plastic will flow uniformly toward the rim (which is desirable); however, there is now the heavy rim to be filled through the narrow flow path through the base, and there will be less pressure to fully pack out the rim. Since the pressure is lower, the plastic in the rim will shrink more than the center section, and, unless it is kept longer in the mold, the product will be deformed after ejection. This is particularly noticeable for plastics with high shrinkage factors.

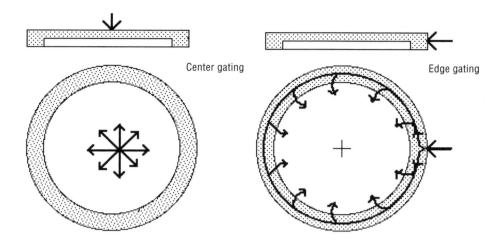

Figure 4.10 Schematic drawings of a shallow, round dish with a thick rim show plastic flow with center gating (left) and with edge gating (right)

    Edge gating is the worst possible flow path: the plastic flow *always* takes the path of least resistance and circles the thinner center before starting to fill it. The plastic will finally converge toward the center and, unless there are provisions in the mold (spot vents) to allow escape of the air trapped by the converging plastic streams, there will be a pronounced mark (a hole, void, burn, or weld) at the point where the plastic finally joins.

    While the examples here just show a simple, round dish, the same principle of the injected plastic selecting the *path of least resistance* applies to practically every product. This problem is not always as apparent as in the examples above, and often will be noticed only after also being overlooked by the mold designer. Eventually, such oversight shows up as an unfilled spot in the product, or as an unacceptable surface blemish. By that time, it is often too late. Although it may be possible to add some vents to the mold, change the gate location, or add more gates, all these mold changes can be very expensive.

    Some typical examples of designs that avoid heavy rims or any other thickening at the end of the flow path are shown below (Fig. 4.11). The end of the side wall, at the rim of any container, whether a drinking cup or industrial container, should for practical reasons be stronger (stiffer) than the wall itself. From design practices for other materials, typically metals, the first thought of the designer would be to increase the thickness at this spot. But with plastic, uniformity of thickness should be the first guiding rule; if necessary, narrow ribs (see arrow in Fig. 4.11) can be added to strengthen the rim without increasing the thickness.

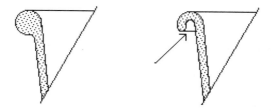

Figure 4.11 Container rims can be made stronger (stiffer) than the container wall by increasing the thickness of the rim (left) or by adding a rib without increasing thickness (right)

The thickening of the rim in the left of Fig. 4.11 is not only a waste of plastic; it will take longer to cool and will shrink more than the rest of the product. It may also show voids. The rim on the right is almost as strong, without these drawbacks.

Designers should follow this basic principle: Do not feed heavy sections through narrow walls. This principle applies to any design, not only to containers.

The filling speed of plastic (in a cavity), in other words, the volume Q that passes through the gap H depends essentially on three variables: the pressure drop P_x, the viscosity μ, and the free space H between the surface of the cavity and the core. This can be expressed as follows:

$$Q = \frac{P_x H^3}{12\mu}.$$

All other things equal, the flow is (theoretically) proportional to the third power of the distance H where the plastic must squeeze through. Doubling the gap (or the thickness of the product) will permit $2^3 = 8$ times the amount Q of plastic to pass through the gap defined by H. In practice it may be less, depending on other circumstances, but it is more than just double the quantity passing through the smaller gap. This relationship shows why the plastic flow prefers the path of least resistance.

4.4.1.4   Parting Line

Rule 4: The outside surface should always end at the parting line.

The "parting line" (P/L), which is actually a parting "plane," is the surface (area) where the cavity and core meets, and where they are clamped together under the locking force of the clamping mechanism of the molding machine.

Designers must understand that, whether the cavity is gated at the edge or near the center of the product, most or all of the plastic will flow toward the parting line. The air that was inside the cavity space before the mold was clamped up must now escape to 1) permit easy filling of the cavity, 2) prevent burning of the leading edge of the plastic, or 3) eliminate the possibility of unfilled spots forming in the product.

Venting at the parting line is no problem; therefore, a surface ending there can always be well vented using simple methods well known to the mold designer. (Unfortunately, the same cannot be said about venting for ribs, hubs, etc.; this will be covered later.)

The shape of the product, at the parting line, needs special considerations:

1. What shape is required for the product? A sharp edge? A rounded edge?
2. Which shape is easier to produce when building the mold?

In Fig. 4.12A, the wall surface ends with a sharp corner (S) on the outside (at the parting line) and a radius (R) on the inside. This is a "natural" shape. The corner S is sharp because two mold parts meet there. The radius R represents the cutting corner of the milling, turning, or grinding tool used to produce this corner. For increased core strength (to reduce the notch effect and the resultant

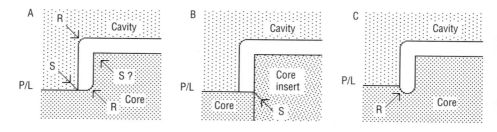

Figure 4.12 Three different product shapes at the outside edges: A. sharp outer edge with round (radius) inner edge, B. sharp corners on both inner and outer edges, and C. Round edges (radius on both inner and outer edges)

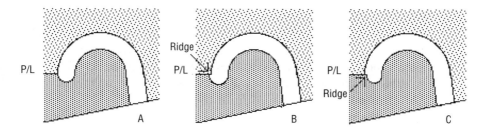

Figure 4.13 Parting line for a rounded edge: A. ideal, B. and C. mismatched

weakness of the core), the radius R should be as large as possible. It is not practical to specify a corner radius smaller than 0.25 mm (0.010 in.). With this shape (Fig. 4.12A), there is no problem with mismatch between cavity and core.

The radius (R) on the inside of the cavity shown in Fig. 4.12 should also be as large as possible, partly for the same (strength) reasons, but also because a round corner will improve the molding condition by allowing easier flow around a radius. In contrast, a sharp corner increases the thickness of the product without adding to its strength. The sharp corner at the top of the core (S?) is easy to make but is not a good shape for the plastic flow. We will look at this particular area later on.

In Fig. 4.12B, the design requires a sharp corner (S) in the core where indicated by the arrow. This may require the placing of a core insert (at extra cost) so that a natural sharp corner (S) is created between the core and the insert. Otherwise, the corners are similar to those in Fig. 4.12A.

In Fig. 4.12C, the design requires a rounded edge at the end of the side wall, so the parting line must be shifted to where the radius meets the outside wall created by the cavity. There is usually no particular reasoning against this design, except that it is somewhat more expensive to produce. For very light products, especially when stripper ejection is used, the radius that is cut into the stripper might cause the product to hang up and not fall free from the mold. Special mold features such as air blow off or mold wipers may then be required to ensure that all pieces have ejected. There is usually no problem with heavier molded pieces weighing maybe 20 grams or more.

The main problem with a rounded edge at the end of the wall is that mismatch is possible between the cavity and core which can result in a small but sharp corner (ridge) where it might not be acceptable for the product (Fig. 4.13). An example is the rim of a drinking cup, where a ridge could be disagreeable on the lips when drinking from such a cup.

There are several methods for avoiding or hiding the mismatch:

1. Reduce the significant tolerances of cores and cavities, and on the dimensions affecting the alignment (elements) of the mold. This can be *very expensive* but can (and should) be done with high quality mass production molds to ensure that all products are as close as possible to the ideal shape and that the mold components are interchangeable.

2. Correct the mold after it is completed and tested: the mismatching areas are blended by hand, by grinding the offending ridges. This is frequently done but not recommended because, in multicavity molds or where a number of identical molds are used for the same product, each product, from each cavity, may be different (see Fig. 4.14).

3. Decide on which side a ridge would have the least disagreeable effect on the end user or on the performance of the product, and then dimension one side different from the other so that, in the worst possible case (all tolerances in the "bad" direction), the piece still will be acceptable for its intended use. In this case (Fig. 4.15), the half-round recess in the stripper is intentionally slightly smaller than the thickness of the rim. With all tolerances in the "ideal" direction,

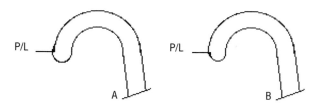

Figure 4.14 Mismatch may result in each product from identical cavities (A and B) being different

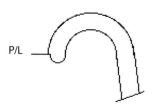

Figure 4.15 The step created in the ridge as a result of worst tolerances still produces a useable product

there will be a minimal ridge. With all tolerances adding in the bad direction, the step of the ridge will be the sum of the tolerances but still is acceptable. In the case of a round container, as illustrated, the dimensions affected are the diameters of the rim in the cavity, the core, and/or the stripper ring.

Similar methods are used to avoid any parting line mismatch, regardless of the shape of the parting line: round, oval, rectangular, irregular, or offset. The more difficult challenge is to foresee where mismatch can occur, and to determine how to dimension the product.

4.4.1.5   Parting Line Selection

Rule 5: Wherever possible, design a product so that the parting line can be in one plane.

In most molds, the P/L is at right angles to the direction of the opening of the mold, as was shown in the foregoing examples; this is the simplest and best condition. For a good mold, the parting line should be ground on both the matching faces of the cavity and the core to avoid any "flashing" (escaping plastic film or burr) of the mold during injection. Such flash must be removed after molding and can require very costly (manual or mechanical) operations; therefore, it is least expensive to provide a well-fitting parting line by grinding.

With some products, it may be necessary to have a parting line which is at an angle to the direction of the mold. If this is unavoidable, it is still not to too expensive as long as the P/L is in one plane. In the example shown in Fig. 4.16,

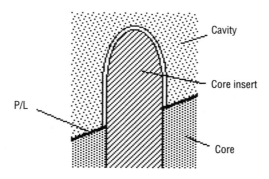

Figure 4.16 A core insert on an angled P/L prevents interference by a projecting core with grinding

the core insert can be removed for grinding. If this was not the case, it would be more difficult to grind this P/L because the core projecting above the P/L would interfere with the grinding wheel.

In the earlier days of mold making, an offset or inclined parting line was often finished by hand, using the method of "bluing and scraping" of the matching faces until a good fit was achieved. This is slow, costly, and never as good as grinding, but it may have to be used even today if there is no other way.

In some instances, electric discharge machining (EDM) is used. After one face is finished, it is used as the electrode to finish the matching surface. This, too, is slow, expensive, and not always possible.

4.4.1.5.1 Venting at the Parting Line
The venting at the parting line is usually easy; to achieve a uniform vent gap large enough to let the air escape but small enough to keep the plastic from flashing, vents are usually produced by careful grinding after the P/L is finished. The depth of a vent is 0.005 to 0.015 mm (0.0002 to 0.0006 in.), and occasionally somewhat larger, depending on the location of the vent and the viscosity of the injected plastic.

The product designer who wants more information on venting is referred to *Mold Engineering* [1], also by this author.

If the parting line is offset, which may be necessary for certain products such as cutlery (e.g., the shape of the spoon or fork handles), the grinding of the parting line and of the vents becomes much more difficult and expensive, especially if the core projects above the P/L. The offset parting line shown in Fig. 4.17 seems simple, but it requires very accurate grinding of all planes indicated by arrows, on both the cavity and the core, to ensure "perfect" fit.

In summary, while it is possible to make any reasonable shape of parting line, the cost of any P/L not at right angles to the opening of the mold, or of any offset, will be much higher. *This is an important point that should not be overlooked during product design.*

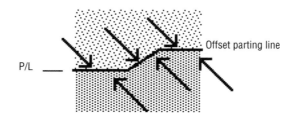

Figure 4.17 An offset parting line has many planes

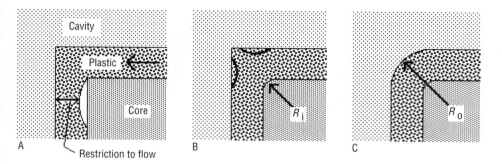

Figure 4.18 Three corner shapes: A. restriction to flow at a sharp corner, B. Shrinkage and sink at a corner with radius R_i, and C. rounding of both the inner and outer corner

4.4.1.6 Changes in Flow Direction

Figure 4.12 on p. 42 shows a typical change of direction in the plastic flow as pointed out by arrow "*S?*" Such changes in direction occur frequently and must be carefully considered before finalizing a design.

A sharp corner should be always avoided, for several reasons. In Fig. 4.18, three different corner shapes are shown. The sharp corner in Fig. 4.18A has no radius, so the plastic will flow over the corner and create a restriction behind the corner, which creates a pressure drop, affecting the filling of the mold further downstream. The void shown next to the restriction will disappear as soon as the mold is filled and the plastic pressure in the mold rises.

In Fig. 4.18B, there is a radius R_i on the core; this radius should be large enough to permit easy flow of the plastic. However, the result of such a design is a thickening at the corner which can result in more shrinkage at this point, creating sink marks as indicated by heavy lines. Also, unless the product really requires a corner shape such as that shown in Fig. 4.18B, the material there is wasted. In addition, sharp inside corners in any piece made from steel (or any other material, including plastics) are bad for the life of the product (or mold part) due to the notch effect created by such an inside corner.

A minimum inside radius R_i is suggested to be between 0.5 and 1.0 times the wall thickness t, but this may not always be possible. See Fig. 4.19 and Table 4.1.

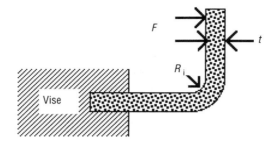

Figure 4.19 A strip with thickness t and a corner of radius R_i is held while force F acts against the inner wall

Table 4.1 Stress Levels as a Ratio of Corner Radius to Wall Thickness

Ratio of r/t	Stress level
1.5	1.2
1.0	1.3
0.75	1.4
0.5	1.6
0.25	2.0
0.1	3.2

The strip with thickness t in Fig. 4.19 is solidly held in a vise. When a force F acts in the direction shown, the corner will be stressed. Table 4.1 shows the stress level in the corner. The stress level increases dramatically as the corner radius r gets smaller in relation to the thickness t. On the other hand, increasing the radius r to more than 1.0 x t does not significantly increase the strength of the corner.

Figure 4.18C shows the best corner design,, which has the same inside radius R_i as in Fig. 4.18B, plus an outside radius R_o. The flow path in Fig. 4.18C is the best of the designs shown; however, the designer must be aware that every change in direction of any flow will result in some unavoidable pressure drop.

The product in Fig. 4.18C also has a constant thickness. For uniform wall thickness, the radius R_o should have the same center as R_i:

$$R_o = R_i + t. \tag{4.1}$$

4.4.1.7   Economic Considerations

It is important that the product designer is always aware that the goal is not only to arrive at a product design that will function as stipulated but also that it can be made in the most economical way. In general, virtually every shape can be injection molded successfully. The question is, then, how can such shape be designed without the need for extraordinary measures when building the mold and when running the mold?

The product designer should understand that when the production requirements are low, a relatively large portion of the product cost will be affected by the cost of the mold; therefore, use of a complicated and expensive mold should be avoided, if possible. On the other hand, for very large production, the cost of the mold has very little effect on the total product cost, and even a complicated mold, with all the features required to do the best job, will be the lowest cost solution.

Therefore, it is of utmost importance for the product designer to make sure that the shape of the product will be suited for the best type and quality of the mold selected. This can easily be done where the product and the mold designer work closely together before settling on a new design, or even on a revision to an older design.

Typically, the selection of gate location and the decision between cold runner molds (two-plate or three-plate molds) or hot runner molds are important to the product design when considering the mold cost and the plastic flow within the mold. Today, about 90% of all molds built are still two-plate molds, which are considerably cheaper than three-plate or hot runner molds. For many reasons, in most cases, hot runner molds are better, but these questions must always be asked:

· If I specify my gate in a certain location, how will the mold function, and how much will it cost?
· Is a more expensive mold necessary, or can production proceed just as well with a lower cost mold?

We have already spent quite a bit of time with what may appear to be unnecessary "details." However, the designer must be aware that, once the idea of what to design has jelled, the remaining time before the design can be completed must be used to consider all these details.

4.4.2   Recesses

Recesses can be defined as depressions in the plastic layer. Their defining length and width dimensions are usually larger than their height, and they do not, as a rule, present moldmaking problems. However, recesses can cause a restriction in the plastic flow that may split the flow path and result in backflow, as illustrated in Fig. 4.10 on p. 40. Such backflow may cause flow and/or weld marks, and possibly may even trap air and cause holes in the surface. In fact, wherever air is likely to be trapped, the mold will require special spot vents (which could result in a witness mark on the surface of the product) and may require special molding conditions (higher temperatures and pressures, and longer cycle times), as well.

   Such recesses could be eliminated from the design by reshaping the wall to maintain uniform thickness. If this is not possible for functional reasons or because of appearance, little can be done except to provide ample radii to facilitate the plastic flow. The product will probably require a slower molding cycle because of the special molding conditions required.

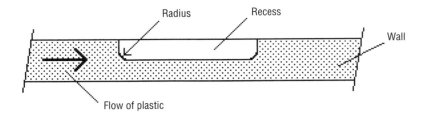

Figure 4.20 Sufficient radius at the base of the recess facilitates the plastic flow

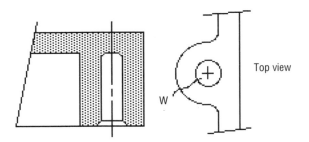

Figure 4.21 Poor plastic flow inside a hub results in a poorly located weld line (W)

The inside of a hub is a special type of recess. It is important to consider how the plastic will flow to form the hub. Depending on the dimensions of the hub and the side wall or rib, and on the location of the gate, the flow may split and converge again at a weld line (W). In Fig. 4.21, the weld line is in a poor location because it weakens the plastic there; if, for example, the recess is used to receive a screw and then is stressed, it may burst along the weld line.

4.4.3   Holes and Openings

Holes are usually round openings in a bottom, side wall, or rib; they can be small or large. The effect of any hole on molding is that the plastic flow will split and flow around the core (or core pin) creating the hole. Where the flowing plastic rejoins, there will be a line, not necessarily visible, called the "weld line" (Fig. 4.22).

Weld lines can cause 1) a physical weakening, because the joint at the weld is usually weaker than the homogeneous, uninterrupted material; and 2) a poor surface appearance. The appearance of weld lines can usually be improved or completely eliminated by changing molding conditions, such as increasing temperatures and pressures and lengthening molding cycles, which will add to the cost of the product.

The product designer, however, can also prevent weld lines or reduce their effect by anticipating the flow of plastic by relying on the designer's own experience or by using flow analyses for different shapes and conditions. The designer can specify the location of one or several gates, and select the shape of the product (wall thicknesses, flow paths, radii, etc.) so that weld line(s) will occur where they will cause the fewest problems, both in strength and in appearance.

(NOTE: For a shape such as that shown in Fig. 4.22, a weld line could be completely eliminated by gating inside the hole using a special method of gating

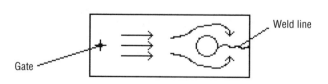

Figure 4.22 Plastic flowing from the gate splits and flows around a hole, meeting at the weld line

called the (cold runner) diaphragm gate. However, this method is rarely used because it increases the cost of the product.)

4.4.3.1   Holes in the Bottom (or Top) of the Product

There is usually no problem in creating a hole in any portion of the product that is at right angles to the direction of opening of the mold. Small holes or openings are often made with core pins; larger holes and openings can be machined right out of the core (or the cavity).

The advantage of core pins is that they can be easily replaced if damaged. The disadvantage is that they must pass through larger parts of the cavity or core where they are located, thereby hindering the cooling layout of these parts. Another disadvantage is that core pins are not well cooled unless provided with their own, internal cooling, which is more expensive.

Where the cores for the holes are left standing on the mold part, their cooling is almost as good as the mold part itself. Damage to such cores, however, requires either welding or replacement of the damaged part.

Why stress this subject, which is really a moldmaker's problem? When specifying small openings, the designer should think of how they are going to be made. As a crass example: Imagine a sieve (or a filter) which requires many holes (Fig.4.23A). If these holes are specified as round, there are two possible methods for creating them:

1. Create the holes using many small projecting pins (Fig. 4.23B), passing though the core (or the cavity), which would make it virtually impossible to place any cooling lines there. This is not a good idea if high productivity is expected. (Note that this method is quite easy in metal stamping dies.)
2. The round core pin (studs) could be left standing from the steel (Fig. 4.23C) of the core (or cavity). This would be good for high productivity, since the mold part could be well cooled, but the steel

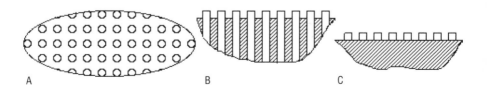

A   B   C

Figure 4.23 Round holes in a sieve (A) may be created using B. small projecting pins passing through the core or C. core pin studs.

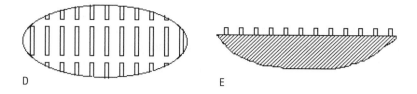

Figure 4.24 Rectangular slots in a sieve (D) are created by milling or grinding the steel core or cavity (E)

around the small studs would have to be removed; such a core would be very difficult (but not impossible) to machine, and subsequently to polish, if the spacing between these studs is very close.

For the same example, the designer could decide that the product would work just as well for its intended purpose if the holes were square, rectangular, or triangular, shapes which could easily be generated by milling or grinding the top of the core (or cavity). Rectangular slots as shown in Fig. 4.24 could be even smaller than would be practical with round holes. This shape would then allow an easy method of manufacture (milling, grinding, and polishing) and result in a solid mold part which can be well cooled.

4.4.3.2   Creating Holes (or Openings) in Product Walls

The problems associated with creating holes or openings at right angles (or any angle) to the mold opening motion are quite different from those previously described for openings made using only the simple (opening) motion of the mold. The cores which create the openings from the side present an obstacle to product ejection. There are different methods which, accordingly, will affect the product cost either by increasing the mold cost, or increasing product cost by slowing the mold cycles or requiring post-molding operations.

The designer must not forget the main goal: the product must function as required but also must be produced at the lowest cost. Where low production quantities are expected, therefore, the manufacture may be planned so that some side openings are not molded but instead are drilled or stamped after molding. The savings in mold cost can easily outweigh the added cost of machining. A decision to select this method can also affect the shape and size of the opening. The product drawing must indicate that the opening will be added after molding, and must describe how this should be done. For mass production, however, the designer should endeavor to mold all openings.

4.4.3.2.1 Side Cores   For openings created using side cores, the wall can be vertical (as shown in Fig. 4.25) or at some (small or considerable) angle to the vertical. Any size and shape opening can be produced.

The resulting opening can have sharp, chamfered, or rounded edges on the outside of the wall (Fig. 4.25A–C). The inside, where the side core meets the main core, will always be sharp. This method is frequently used and is based on either cam-actuated or hydraulically operated slides which move the side core in and out of molding position. This motion can be at right angles to the center of the vertical axis of the product, or at any (reasonable) required angle, in which case the mold cost would increase even more.

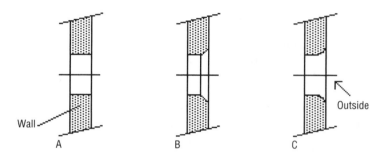

Figure 4.25 Opening shapes created by side cores: A. sharp edges, B. chamfered edges, and C. rounded edges

4.4.3.2.2 Moving Side Walls   A moving side wall is a section of the cavity which forms the side (or a portion of the side) of the product. The section either moves straight sideways, similar to a side core, or it slides up (together with the withdrawing core and the product) and outward at the same time. This compound motion (up and out) frees the side wall from any projections that are formed by that section of the cavity and from any cores that have openings.

Figure 4.26 depicts the basic principle of a mold with a side section, which in this case creates an opening and two projections. (For clarity, mold details such as cooling, venting, and the method of guiding and actuating the motion of the side section are not shown, nor are the ejectors or strippers that remove the product from the core after the mold has fully opened.)

The main reason for showing this example is so the designer can see how openings and projections in a side wall can be created, and where draft angles, radii, and/or chamfers are required to be shown on the product drawing. If these

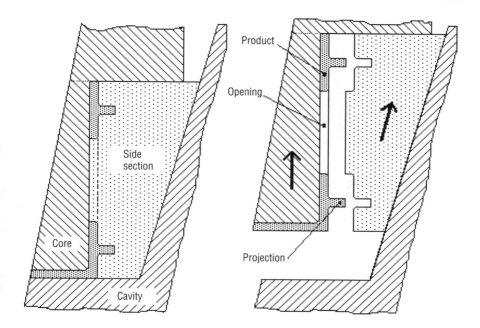

Figure 4.26 Schematic of a basic mold with a moving side wall section shows movement of various mold sections

features are not shown, the mold designer will surely come back and ask for changes to facilitate molding and moldmaking. There would be one side section for each of the side walls of the product, and each section must withdraw to free the product for ejection. Typically, a beverage crate will require four such sections, while a radio cabinet may require only one. The designer can understand how complicated and expensive such molds can be.

4.4.3.2.3 "Vertical" Shutoffs

If the product requires openings only in the side wall, there is a simpler and less expensive method available to create such openings without side cores or moving side sections. The only condition is that the walls must have ample side draft, in the order of 5% or more, as shown in Fig. 4.27 on p. 56.

The opening in Fig. 4.27 is created by raised "pads" on both the cavity and the core that meet at the "vertical" shutoff. As the mold opens straight up (as the arrow indicates) the pad on the core will lift off the pad on the cavity at the P/L indicated; the core will carry the product along with it. The product will slide off easily when ejected from the core, since there are no "hooks" to hold it in place.

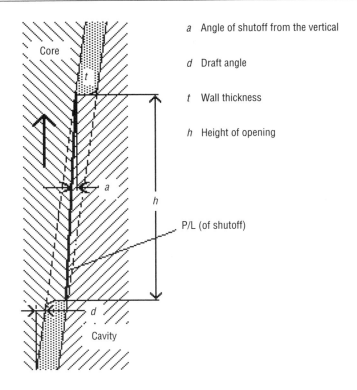

Figure 4.27 A vertical shutoff creates an opening in a tapered wall

It is important that the angle *a* of the P/L where the pads meet is greater than 1% from the vertical so that the mold steels of the cavity and core do not rub against each other during opening and closing of the mold.

The geometry of such openings produced by shutoffs is important. Both angles *a* and *d*, and the wall thickness *t*, will determine the minimum height *h* of the opening in the vertical direction. This relationship must be calculated or laid out in a large-scale drawing to ensure that the selected height of the opening is feasible with the selected dimensions *a*, *d*, and *t*, without creating a hook which could trap the product. Note the location of possible radii at the core and the cavity. This method is often used with stiff plastics.

4.4.3.2.4 Shutoff at the Cavity Wall   There is an even simpler method used to create openings in the side wall, as shown in Fig. 4.28. In this arrangement, the shutoff where the core and the cavity meet to create the opening is located on the cavity wall without the need for pads, as in the previous example (Fig. 4.27). The

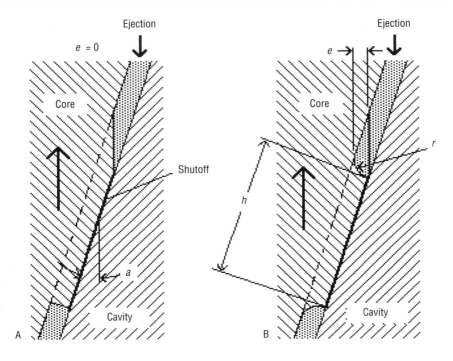

Figure 4.28 Shutoff at the cavity wall: A. simple ejection but unappealing shape, and B. more attractive shape but requires elastic deformation of the plastic

greater the angle of the side wall with the vertical, the easier it is to create the shutoff.

In Fig. 4.28A, as the mold opens, the core together with the product withdraws from the cavity in the direction of the arrow. The product can then be simply ejected (in the direction of the short arrow) without meeting any resistance from the cores which generate the side openings. The shape of the opening edge is not particularly appealing, however, with the sharp edge on top and a plain corner at the bottom. This method will work with any plastic, soft or hard.

Figure 4.28B is a variation of this method. Radii r are shown on both ends of the opening, which is now more appealing. There is no problem with the lower radius, but the upper radius will now interfere with ejection. The plastic must stretch over the hump created by the radius r by the distance e. (The greater r, the smaller e, and the smaller is the resistance against ejection, due to the stretching of the plastic. In Fig. 4.28A, $e = 0$.) Not every plastic may permit being stretched as far as desirable.

The inversion of this principle is also possible, with the shape of the outside of the sidewall cut into the cavity, and the cores which create the opening left standing so that they will "shut off" on the side of the (main) core. The advantage of this arrangement is that the radii can be on the outside of the product, but the inside would then have sharp corners. This method is generally not desired, for two reasons:

1. It is more difficult to machine the side walls on the *inside* of the cavity than on the *outside* of the core; and
2. after withdrawal, the product remains in the cavity, so an ejection system is required on the cavity side which is usually more costly to provide.

4.4.3.2.5 Holes in a Vertical Side Wall

Figure 4.29 shows a very simplified view of a box with a hole in the end wall. The hole is produced by a side core, and some mechanism is thus required to pull the side core out of the wall of the box before the box can pull out of the cavity. This causes a slight increase in the mold opening time.

In Fig. 4.30, the box has been redesigned. To eliminate the side core, the shape of the wall has been modified with a vertical groove to permit an insert to shut off against a matching wall of the core.

The design in Fig. 4.30 has created a composite hole, where the walls forming the circumference of the hole are somewhat smaller than those in Fig. 4.29.; Also, the wall of the hole is divided into two half-cylinders, offset at the shutoff plane. Provided this change is acceptable for the intended use of the product, the mold

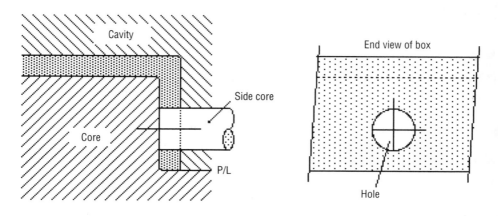

Figure 4.29 Mold schematic (left) shows the side core passing through the cavity to produce (right) a hole in the end wall of a box

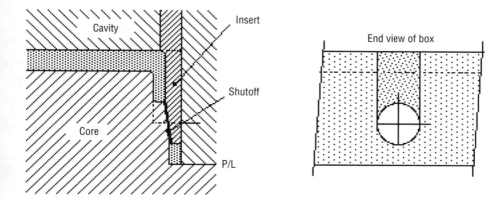

Figure 4.30 Mold schematic (left) shows the insert passing through the cavity to shut off against the core to create (right) a hole in the end wall of a box

becomes much simpler: the cost of the insert is much less than the cost of the side core plus mechanism. Also, the mold can cycle faster. This was a typical example of how a slight design change can affect the productivity of a mold, as well as the mold cost, and eventually, the product cost.

Note that the illustrations in Figs. 4.29 and 4.30 are only schematic; draft angles and mold features have been omitted for clarity. The angle of the shutoff is shown exaggerated. An angle of 3–5% is usually sufficient.

4.4.3.2.6 Holes in a Vertical Rib
A condition very similar to that for holes in a vertical side wall exists when molding a hole in the side of a rib. The use of side cores to produce such a hole is often not practical or may be impossible without incurring major costs.

A mechanism to create such a hole could be placed inside the core, similar to the sliding ejector shown in Fig. 4.53 on page 84. Instead of the hook in that example, the slide could carry a core pin to create the hole or opening. The disadvantages of this system were explained earlier: it would not only make the mold very complicated and expensive but also would make it difficult to place optimal cooling lines for efficient operation.

As shown in Fig. 4.31, an overlapping insert must be accommodated by a slot that allows the insert to pass through the wall. Such slots are usually acceptable and are quite commonly used, especially in technical applications.

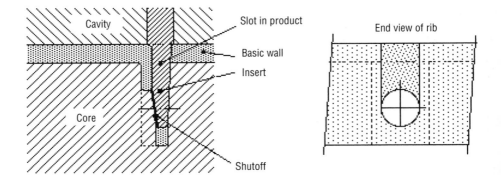

Figure 4.31 Mold schematic (left) shows the insert passing through a slot in the product to shut off against the core to create (right) a hole in a vertical rib

4.4.4   Ribs and Other Projections

The parting line is not always at the apparent end of the outer surface, or wall. For a shape such as the basket or container in Fig. 4.4 on p. 34, the parting line is at the well-defined end of the plastic flow. However, if a container has a shape such as that shown in Fig. 4.32, the outer surface actually continues into the flange and down to the P/L, where it ends. The portion of the wall that extends toward a "dead end" is a "projection" and is, therefore, subject to all the same molding (and design) problems as any type of projection from the main flow of the plastic.

In Fig. 4.32, the projection is in line with some of the wall. In most cases, however, projections are at right angles to the wall. Examples of typical projections are:

- Ribs and gussets (reinforcements),
- studs (usually round),
- mounting supports (any shape),
- hubs (usually round),
- ejector pin extensions,
- flanges,
- threads, and
- undercuts.

As we discussed earlier, any change in plastic thickness has several disadvantages, related to the flow and to the creation of stress risers, which, unless very smooth, can severely affect the product strength in the area of the thicken-

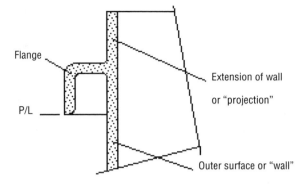

Figure 4.32 The outer surface extends or "projects" beyond the flange of this container

ing. The main problem connected with any projections is that they are accompanied by a thickening of the plastic, with the following, sometimes costly, results:

1. Greater material concentration,
2. longer cooling cycles, and
3. less homogeneity of the plastic; therefore, less strength.

4.4.4.1 Reinforcements (Ribs and Gussets)

Ribs or gussets are used to reinforce or stiffen an otherwise weak or flexing structure. The method of calculation for the moment of inertia and section modulus are basically the same as that for any other machine or structural design, and all formulae used for strength of materials (shown in any engineering handbook) can be applied, except that, for plastics in general, the values for the modulus of elasticity (E) and the values for tensile strength and other properties, are much lower than those for engineering materials such as steel, aluminum, etc. These values can be found in data sheets for the various plastics.

The designer will start with an assumption of the plastic thickness suitable for the intended shape. If precedents exist, this is usually no problem. However, when starting a design "from scratch," it may be difficult to determine an optimum thickness for a shape which has never been molded before. Because the total cost of the molded product depends to a large extent on the mass of plastic used, and also because the thickness will affect the molding cycle, the designer will start with a thickness he or she believes suitable to satisfy these two concerns. Of course, this may result in a product that is not strong enough to satisfy its strength requirements.

In some cases, a handmade mock-up machined from the solid or an actual molding from an experimental mold can provide a fairly good idea of the expected strength and will indicate where reinforcements are required. The other advantage of a prototype is that the function of the product can be better evaluated than from a drawing only. This may be quite expensive, but is often preferable to building a production mold and only then finding out where the weak spots are.

The modern alternative to a mock-up is to have a computer *finite elements analysis* done of either the whole product or the critical areas only. This, too, can be quite expensive, but will be unavoidable with larger products for which an experimental mold or handmade mockup may be not practical. There are professional services and computer programs available for finite elements analyses.

An experienced designer may approach the problem of strength by calculating only those areas which are heavily stressed, using methods similar to those used for machine or structural design, and otherwise going by "gut feel." This method is frequently used but may lead to "erring on the safe side" and overdesigning, which often yields a product that is too heavy or too complicated.

Another possibility is to solely depend on the experience of the designer (or groups of designers) who may have had experience with similar products. This may work in many cases, but there is still the risk of overdesigning and ending up with a product which, on second thought, could have been made lighter or less complicated, and therefore less costly.

The designer usually has two alternatives: either make the plastic thicker (heavier) until the required strength or stiffness is obtained (which is expensive), or add ribs or gussets in strategic locations where they achieve the same result as adding plastic all over but without adding much mass when compared to the originally planned product, where the wall was too thin and the product not strong enough.

Of course, there may be the occasion where both ribs or gussets *and* thickening of the bottom and/or sides will be unavoidable. But the designer will then understand that the piece will mold slower and cost more in production.

4.4.4.2   Venting

Before going into more detail, it is important that the designer is aware of the problems associated with venting in a mold. We have already discussed venting in the context of parting line (p. 42), but whenever there is a rib or a hub, etc., the designer must anticipate the expected flow of the plastic in the mold to fill it out. It is up to the *mold* designer to provide adequate venting for any dead ends into which the plastic will be driven during injection, trapping any air present. Such venting can sometimes increase the mold cost substantially, and may even affect the cooling of the mold because venting provisions may interfere with a good cooling layout, resulting in slower molding cycles and decreased productivity.

Wherever there is a deeper rib or a hole, there must be an ejector pin or ejector sleeve which, by virtue of the required sliding fits, acts as a "natural" and "self-cleaning" vent and, similar to the parting line vent, is a desirable feature in a mold. If there are no ejectors planned, other methods of venting will be required ("stationary" or "fixed" vents) which are not self-cleaning and which could require frequent maintenance.

4.4.4.3   Shrinkage

Before continuing, we will refer to **Rule 1**, stated earlier on p. 35, and provide a related statement about uniform thickness:

> **Rule 6: It is virtually impossible to avoid wall thickening at projections or ribs.**

Therefore, the designer must expect sinks or voids; these are often acceptable for the product. However, if such "cosmetic" flaws are not acceptable, the molding conditions must be adjusted to prevent the development of unsightly sink marks on the product surface opposite such thickenings, which can be accomplished by using higher injection and hold pressures, longer injection and cooling times, and higher melt and mold temperatures; all of these measures will result in slower cycles.

Plastic in any thickening nearer the end of the flow path will shrink more than any thickening nearer the gate, a factor which is especially noticeable for plastics with high shrinkage factors (see also p.35). Plastic will always take the path of least resistance—from the gate to the areas remote from the gate (**Rule 3**, p. 38). This point is particularly important, because ribs and gussets may allow the plastic to take a different path away from the gate; the fact that there is a rib may entice the plastic to flow faster through the rib and arrive at areas remotest from the gate before the plastic would arrive there if there was no rib. This may create an encirclement of the inner, unvented areas of the mold (see Fig. 4.10, right), and cause similar problems as explained earlier.

Venting, shrinkage, and flow paths affect the design of any projection. Considering the flow path of the plastic, therefore, is of utmost importance in avoiding unpleasant surprises when the mold is sampled.

4.4.4.3.1 Predicting the Expected Flow Path   Unless the designer can rely on his or her own experience, or the help of an experienced mold designer, the designer should utilize a computer program that can simulate the flow path when a suggested design is ready to be analyzed. There are several such programs commercially available, and there are consultants who specialize in this field.

4.4.4.4   Flat Surfaces

Another point which requires careful consideration is the design of "flat" surfaces. Although already mentioned on p. 35, it is important enough to bear repeating: It is *very* difficult to produce a really flat surface, especially in high shrinkage plastics, unless molding conditions are carefully controlled, which unavoidably results in slower molding cycles.

The main obstacle to achieving flatness is uneven shrinkage. Uneven shrinkage can be caused by:

- Uneven thickness at the base of any projection, such as a rib;
- bosses that usually cannot be properly cooled; or
- poorly designed mold cooling.

Most *mold* designers will attempt to provide the best possible cooling in molds, especially if large production is expected. However, the design of cavity and core cooling in areas where projections are located may be hampered if they are too close together, which makes it very difficult to provide adequate cooling at that location. Such poorly cooled areas of the mold will then control (slow down) the molding cycle.

4.4.4.5   Ribs

The left illustration in Fig. 4.33 shows a rib which has (at its base) the same thickness *t* as the base itself. While this design is stronger than the one on the right, the thickening (symbolized by a circle) is greater than that of the sketch on the right. .Such a thickening can cause a sink mark; therefore, the design at the right is usually preferable. Ideally, the thickness of the rib at its base should be only about 0.5 *t*.

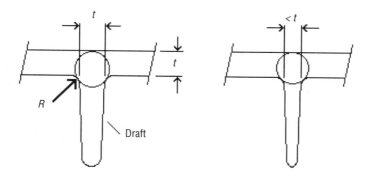

Figure 4.33 Cross sections of two ribs: (Left) rib thickness at base equals base thickness; (right) rib thickness is less than base thickness

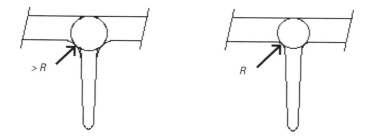

Figure 4.34 Cross sections of two ribs: (Left) large radius, (right) small radius

Note that the sketches in Fig. 4.33 show draft angles. Draft angles will be discussed later, in a separate chapter.

The radius in the corner at the base will also present a problem. According to Table 4.1 on p. 48, the stresses increase rapidly as the radius gets smaller. Here, again, the designer must select a compromise: the larger the radius, the better for the stress conditions, but, at the same time, the thickening will be larger, as shown by the circles in Fig. 4.34. This figure shows the same ribs as before; however, the radius R on the left is larger than that on the right, and therefore the bulk of plastic is larger.

It is up to the designer, in consultation with the client, to decide whether to sacrifice appearance (accept sinks) or to accept a higher product cost as a result of slower production and wasted plastic.

A major problem with rib design is the proper filling (molding) of the ribs. Shallow ribs no higher than $H = 2t$ (Fig. 4.35) do not present any problem, as a rule.

Since flowing plastic will always take the path of least resistance, the plastic will fill the base before entering the rib. As a result, the air in the rib space is trapped. As the plastic continues to fill the space in the cavity or core where the rib is being formed, the trapped air is compressed. This can result in two possibilities (Fig. 4.36): if the plastic enters rapidly and the volume of air is large, the compressed air can heat up so much that the plastic will burn at the crest of the rib. If the plastic enters slowly, the compressed air will form a pocket, and the rib could be unfilled at its crest. In some cases, either flaw may be acceptable, but as a rule neither is acceptable and the product will be rejected.

The problem of trapped air in ribs increases with the height H of the rib (depth of the "groove" in the mold). For higher ribs ($H > 2t$): the volume of entrapped air increases; this air must be removed, or "vented," to ensure proper formation of the rib. Although this is really a mold design problem, the product and mold designers, by working together, can often find a suitable and relatively inexpensive solution to keep the mold cost down and to avoid costly venting solutions.

If the rib ends in an outside wall, the air can usually escape through the parting line vents (Fig. 4.37). However, if the rib does not end at the outside wall,

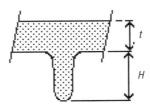

Figure 4.35 A shallow rib cross section showing dimensions H and t

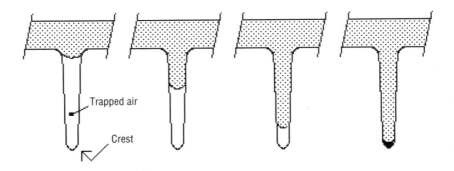

Figure 4.36 Trapped air at the crest of a rib is compressed and may result in unfilled ribs or burned plastic at the crest

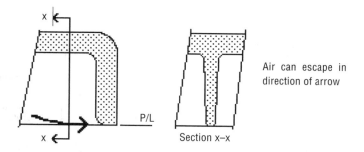

Figure 4.37 Rib ending at parting line

or if the plastic flow coming from the (heavier) wall arrives at the end of the rib before the plastic flow through the rib, the air will be blocked. In such cases, vents must be provided at the point(s) where air entrapment is expected. For a narrow rib, this can present a major mold design problem, especially if the rib is less than 2 mm thick at the crest.

An inexpensive solution is to add ejector pins at the crest of the rib. This not only permits a good, natural venting through the clearance gap between the ejector pin and the core block but also provides the best possible point from which to eject the product from the mold, *if pins are planned for the ejection of the product.* A disadvantage is that such added ejector pins can have a detrimental influence on the mold cooling by limiting the space available for cooling lines, which cannot be located where there are ejector (or any other) pins. The location of the pins can often be selected so that they will provide sufficient (if not perfect) venting without too much effect on the cooling line layout.

Figure 4.38 (left) illustrates an instance where a gusset or rib is required to stiffen a wall, but the rib is too thin to add a reasonably sized ejector pin. Since the rib does not extend to the parting line, the trapped air must be vented. To improve the design, a stud has been added to Fig. 4.38 (right), with a diameter selected to be slightly larger than a standard size ejector pin. The trapped air can now escape through the clearance where the ejector pin passes through the mold.

This arrangement has the added advantage that the rib can be ejected from the lowest point, and it should ensure that the rib will not remain stuck in the mold at ejection. On the down side, the addition of the stud will fractionally add to the molding cycle because of the increased mass and the possibility of a sink mark, but it is better to mold somewhat slower than to be uncertain whether the piece, as designed, can be molded at all.

The example in Fig. 4.39 is given with the intent to show the effect of stiffening of a rib added to a flat surface. For simplicity of the example, a strip

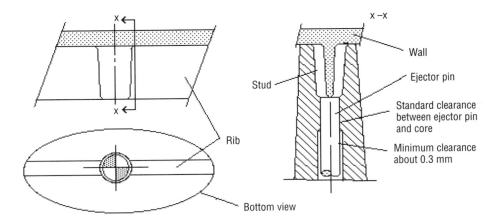

Figure 4.38 Side and bottom views (left) of a rib designed to stiffen the product wall; cross section (right) shows a stud added which is slightly larger than the ejector pin

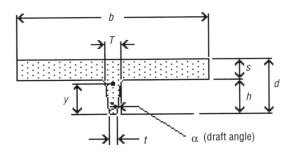

Figure 4.39 Cross section of a rib added to a flat surface shows dimensions for calculating the effect of thickening the flat surface

is assumed to be 1 inch wide ($b = 1"$) and 0.1 inch thick ($s = 0.1"$), and a rib is designed so that it will minimize the effect of thickening the flat surface. The rib thickness was selected to be $^5/_8$ of the thickness s (or $T = 0.0625"$) at the base of the rib, and $^3/_8$ of the thickness s (or $t = 0.0375"$) at the crest of the rib.

The following formulae are shown in any handbook on strength of materials. The area (A) of the cross section of the profile is:

$$A = bs + [(T + t)h]/2 \qquad (in^2) \quad (4.3)$$

The moment of inertia (I) is:

$$I = 0.0833 [4bs^3 + h^3(3t + T)] - A(d - y - s)^2 \qquad (in^4) \quad (4.4)$$

Table 4.2 Relative Stiffness of a Beam with Reinforcement Ribs as per Fig.4.39

h (in)	a (%)	A (in^2)	Increase in weight (%)	y (in)	I (in^4)	Gain in stiffness over flat strip
0 (flat strip)	n/a	0.1	basic weight	0.05	0.000083	basic stiffness
0.1	10.0	0.105	5	0.1454	0.000132	1.6 times
0.2	5.0	0.11	10	0.2371	0.000299	3.6 times
0.3	3.3	0.115	15	0.3255	0.00652	7.8 times
0.4	2.5	0.12	20	0.4111	0.003402	15 times
0.5	2.0	0.125	25	0.4941	0.002154	26 times
0.6	1.7	0.13	30	0.5754	0.003402	41 times
0.7	1.4	0.135	35	0.6538	0.005047	60 times

The distance from the neutral axis to extreme fiber (y) is:

$$y = d - [3bs\,2 + 3ht\,(d + s\,)\,h\,(T - t\,)(h + 3s\,)]/6A \qquad (4.5)$$

By introducing into these formulae the values selected for the reinforced strip shown in Fig. 4.39, we get the results arranged in Table 4.2.

Note that these numbers are an example only and apply to *one specific relationship* of dimensions as shown in Fig. 4.39, depicting a cross section through a beam. The values for A, I, and y must be calculated for each application (shape and set of dimensions).

To demonstrate the significance of the values for I (moment of inertia) and y (distance of neutral axis from extreme fiber), we again refer to some basic formulae, found in every machinery textbook or handbook, listed under "strength of materials", with accompanying figures. A number of different, frequently occurring cases are listed as "case 1," "case 2," etc. We will be looking at only two typical cases which show how the values I and y are used to determine deflection and stresses at the extreme location. The reader who is interested in more details is advised to consult any textbook or handbook on strength of materials.

The values of I and y for the solidly mounted beam in Fig. 4.40 can be calculated from the shape of the cross section of the beam, as was done for Fig. 4.39 in Table 4.2 for eight different cross sections.

The maximum deflection f—*for this case only*—can be calculated from the following formula:

$$f_{max} = \frac{WL^3}{3EI}. \qquad (4.6)$$

The value E is the modulus of elasticity for the material of the beam (in our case, the plastic) from which the product will be made. Note that the product is

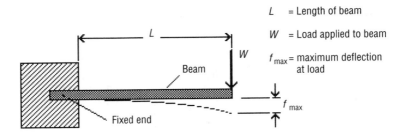

Figure 4.40 Cross section of a solidly mounted beam shows dimensions used to calculate deflection and stress

rarely just a beam but is, in stress analysis, broken up into many (finite) elements (beams) which each can then be calculated for adequate strength. There are a number of specialists and computer programs to perform such "finite elements analyses." We will not further go into this subject.

From equation 4.6, it is seen that the greater I, the less the deflection f. Referring back to Table 4.2, this influence is shown in the column "Gain in stiffness over flat strip;" a flat strip has the least resistance against deflection. The increase in I is quite spectacular as the height of rib increases. However, there are the other factors to consider, such as molding (filling, venting) and difficulties in making the mold with deeper ribs; such factors will usually limit the height of ribs to manageable proportions. Another important limitation to the height of ribs is the stress level at the crest of the rib (the point farthest from the neutral axis). This (or any) point must never be stressed beyond its permissible limit, that is, the tensile strength of the plastic, but must instead be reduced by an adequate margin of safety, or "safety factor," suitable for the performance of the product.

The maximum stress s_{max} in the beam can be calculated—*for this case only*—from the following equation:

$$s_{max} = \frac{WL}{Z} \qquad (4.7)$$

where Z is the *section modulus* and is defined by

$$Z = \frac{I}{y} . \qquad (4.8)$$

By substituting the value of Z into equation 4.7, we get

$$s_{max} = \frac{WLy}{I} . \qquad (4.9)$$

Both the values I and y can be calculated using equations 4.4 and 4.5.

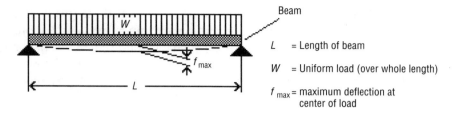

Beam

L = Length of beam

W = Uniform load (over whole length)

f_{max} = maximum deflection at center of load

Figure 4.41 Cross section of a beam supported at each end and under a uniform load

We see that the greater I, the smaller the maximum stress, but the greater y, the greater this stress. In any case, s_{max} must be well below $s_{permissible}$ for a well-designed product.

The illustration in Fig. 4.41 shows another typical case. Here, a beam is supported on both ends and loaded with an evenly distributed load.

The formulae for deflection and stress—*for this case only*—are as follows:

$$f_{max} = \frac{5WL^3}{384\,EI} \qquad (4.10)$$

and

$$s_{max} = \frac{WL}{8Z}. \qquad (4.11)$$

The section modulus Z is the same as calculated in equation (4.6), therefore

$$s_{max} = \frac{WLy}{8I}. \qquad (4.12)$$

Several conclusions can be drawn from Table 4.2:

1. The stiffness of the beam grows very fast as the height of ribs increases.
2. There is an increase of mass in this section of the product. Is the increase in stiffness really worth the added mass (and cost) of the product?
3. For moldmaking, the increase in height means a smaller draft angle a and more difficulty (higher mold cost) in making (polishing) the rib so that it will easily release from the mold.
4. Up to a rib height of about $H = 1\,s$ to $2\,s$, there is usually no problem with filling, even without provision for venting. Any deeper ribs will need venting. If the rib ends at the wall at the parting line, or if there are ejector pins pushing at the crest of the ribs, there are usually

few problems with venting and filling. However, if the ribs do not end at the wall, provisions for venting are imperative, and a more complicated mold construction (special ejectors, vented inserts, or vent pins) may be required and therefore increase the mold cost.

5. It may be of overall advantage to have *several shallower ribs*, more closely spaced, rather than one deep rib.

This is all pointed out so that the product designer will think twice before specifying deep ribs, and if necessary, make a detailed stress analysis of the product before designing ribs which are unnecessarily deep, with all their potential problems and cost implications in molding and moldmaking.

4.4.4.5.1 Using Ejector Pins Under Ribs

A serious problem with ejector pins is that they often present mold maintenance problems, especially if the pins are long (and more so if the product requires a long ejection stroke). Any pin size smaller than 3 mm in diameter (even though they are commercially available in sizes as small as 1.5 mm in diameter) should be avoided to prevent frequent pin breakage and other maintenance problems.

If there are other methods of ejection planned, such as the use of strippers or air ejection, the ejector pins and their mechanism can add considerably to the mold cost. Also, the effect on cooling (as outlined before) will increase the molding cycle and thereby reduce the productivity of the mold.

4.4.4.6   Gussets

Basically, gussets are short ribs used either to 1) stiffen a wall where it makes significant turns, or 2) to support long, slender hubs or studs. A few typical examples of stiffening at changes in direction or at corner are shown below.

4.4.4.6.1 Gussets in a Box

In a box as shown in Fig. 4.42, a gusset would usually not be required except to stiffen the side walls if they are very long. The stiffening effect of the end walls (sides) will often be enough to ensure the required stiffness of the bottom and sides of the box.

4.4.4.6.2 Flat Strip with Upturned End

The stiffening effect of the end wall is missing on a flat strip, and a gusset may be necessary to ensure that the bent-up portion of the strip will retain its attitude under the expected load condition (Fig. 4.43).

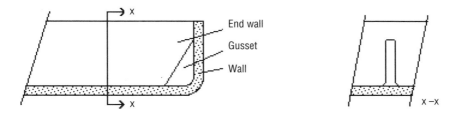

Figure 4.42 Side view (left) of gusset to stiffen end wall and end view (right) of wall

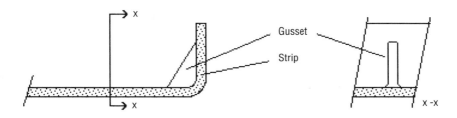

Figure 4.43 Side view (left) of a gusset at a bent region of a flat strip, and end view (right) of the bend region

4.4.4.7   Studs (or Bosses)

Studs do not have to be round; they can have any shape suitable for the design, as long as they can be molded (i.e., withdrawn from the mold part (cavity, core, side core) in which they are formed. Studs are usually intended to:

1. Support other components,
2. locate (and also support) other components,
3. be used as rivets, for later joining other pieces, or
4. permit any other function for which a stud is suitable.

Free standing studs in the proportion shown in Fig. 4.44, or even longer, are rare. Usually they are not more than twice the height of the diameter (or the smallest dimension at the base), because they are very sensitive to side loads from any direction. A small radius where the stud meets the wall is absolutely necessary, but, as discussed earlier, too large a radius will be stronger but will also greatly increase the mass of plastic in this spot.

If there are side loads possible or expected, a design using gussets should be selected. The gussets should be essentially in that plane where the side loads are

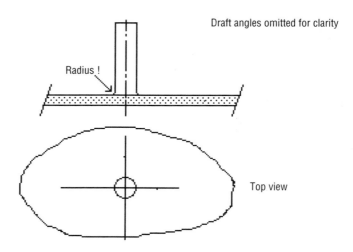

Figure 4.44 Side view (upper) and top view (lower) of a stud

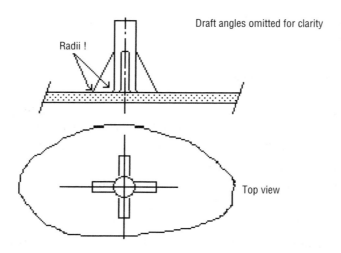

Figure 4.45 Side view (upper) and top view (lower) of a stud with four gussets

expected to act; therefore, in some cases, one or two gussets would be sufficient. Most frequently, four gussets are used, as shown in Section 4.4.4.71, Fig. 4.45. Three gussets, spaced at 120°, could also be suitable, depending on the circumstances.

4.4.4.7.1 Stud with Gussets Preferably, gussets on a stud should be no thicker than ribs for the same base thickness, to reduce the mass of plastic. There is often no need to extend the gussets to the full height of the stud. This not only saves material but also reduces the machining and polishing time of the gusset. On the sides of the gussets, there must be a minimum draft of 0.5°, although 1° or more per side is better, in the direction of the motion of the mold part during ejection or, in case of side cores, in the direction of the motion of the core pull. The stud should also have a draft angle, similar to the suggested angles for the gussets (Fig. 4.45).

The crest (tip) of the stud should be designed so that it is slightly larger than a standard size ejector pin; if this is not the case, this portion of the mold not only becomes more expensive to produce but also the cost of mold maintenance will increase. Thin ejector pins break fairly easily and frequently, and their replacement is easier and less costly if they do not have to be specially made.

In any case, the effect of the thickening of the stud, plus the gussets, will increase the mass of plastic at the base of the stud and, especially with high-shrinkage materials, sinks or voids can be expected if fast cycling of the mold is planned. If voids or sinks are not permissible, the molding cycle must be increased to allow this area to be packed out.

Studs not longer than twice their diameter usually do not require provision for ejection or venting, but they must have a good draft angle (at least 3% per side) and a good finish so that they will pull easily out of the mold, without breaking. The following example shows an application where a larger stud is required to support a large component (usually, there would be several of these supports). To avoid a large mass of plastic or the need to provide a hub (a hollowed out stud), a cross of two ribs could be used (Fig. 4.46).

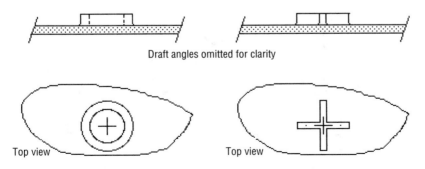

Figure 4.46 Side and top views of a stud (left) and a cross-shaped rib

The examples in Fig. 4.46 show that there are different ways of reaching a similar result (the support or the location of a component), and that it is up to the designer to specify a shape that not only is needed for the required purpose but also can be easily molded.

4.4.4.8   Hubs

A hub can be defined as a hollowed-out stud (or boss). All rules given for studs (height, draft angles, corner radii, size or diameter to be compatible with available ejectors and/or ejector sleeves) apply for hubs.

Hubs are usually used to:

1.   Receive a (usually self-tapping) screw,
2.   match a fitting stud for location of another component,
3.   fulfill some specific function (and therefore shape) in an assembly, or
4.   remove mass inside a heavy stud.

Like a stud, a hub can be free standing, such as the studs illustrated earlier (Fig. 4.44 and 4.45), or it may be attached to a side wall or corner, or be located where ribs meet, as illustrated in Fig. 4.47. The arrows indicate possible sink marks (exaggerated) caused by the heavier section of the plastic in these areas. Side walls and corners provide the best possible stability for the hub.

On its inside, the hub may have a through-hole or a hole ending at the bottom. The hole inside the hub can be round or any other shape required for the design.

If a hub such as that shown in Fig. 4.47A is used in assembly to receive a stud similarly located on top of a rib in the matching product, the designer must be aware that the beneficial effect of the coring-out by the core pin (to reduce the mass) is not available here, and the mass under the stud will be larger than that

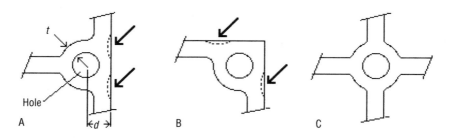

Figure 4.47   Three different locations of a hub: A. at a side wall, B. at a corner, and C. where ribs meet; arrows show possible sink marks where plastic thickness is heavy

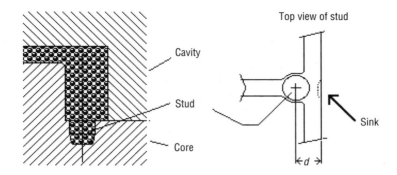

Figure 4.48 Cross section of mold (left) shows the molding of a stud corresponding to the hub in Fig. 4.47A; top view (right) of the stud shows dimension d and location of possible sink mark

of the rib. In Fig. 4.48, the support under the stud is shown as small as practical, to fit the corresponding hub; dimension d and the hole and stud diameters must be in accordance with the proper range of fits selected for this assembly.

NOTE: See also Fig. 4.21 (Section 4.4.2, p. 50). There is always the risk of weld lines, which may weaken the wall of the hub. Flow analysis may predict such areas of weakening.

Even though the hole depth needs to be only slightly more than the height of the stud, it is better to extend the hole in the hub as deep as possible (as shown in Fig.4.49) to reduce the mass of the plastic and to achieve better cooling in this area. The designer should make sure that the core pin, in other words, the mold component that will produce the inside shape, can be easily made, and, if at all possible, that it can be well cooled. The outside of the hub (or stud, or any projection) is *inside* a larger mold component that is usually well cooled. At the circumference of the stud or hub, the heat from the injected plastic can be fairly well dissipated into the surrounding mold part and should not seriously affect the molding cycle. However, the core pin that forms the inside of the stud often has only a small mass. During molding, it is surrounded by hot plastic. Because of its small mass, the core pin will quickly reach the temperature of the plastic. Because of the small cross section of the pin where it is located in the mold (cavity, core, or side core), this heat will dissipate slowly into the mold steel. Therefore, unless the pin is well-cooled, it can dramatically affect the molding cycle.

The arrows in Fig. 4.49 (p. 78) symbolize the heat flow into, through, and out of the core pin into the cooled mold. There are many ways to internally cool a core pin (often at a relatively high cost); the smaller the size of the pin, the more

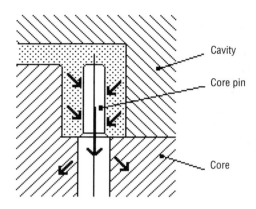

Figure 4.49 A core pin extends deeply into the hub to reduce plastic mass and enable better cooling

difficult (or even impossible) it becomes to provide good cooling. A poorly cooled or an uncooled core pin will heat up during each molding cycle to such an extent that the plastic will not cool down fast enough around the pin, and if the product is ejected too soon, the still hot and molten plastic will stick to the pin and deform the shape of the hole. The molding cycle will then have to be lengthened to give the core pin time to cool down again, before the next cycle can start. (Occasionally, during the mold open period, air is blown at the pins from the outside to cool them, but this too is at a cost.) It is very important that the product designer understands the effect of size and shape of holes when concerned with the productivity of the mold to be built for the product under consideration.

It cannot be stressed enough that the primary goal of the product designer is not only to arrive at a design that is feasible and that will fulfill the stipulated requirements but also to ensure that the product can be made in the most economical way. For this reason, much emphasis is placed on the importance of the mold (and the mold cooling).

4.4.4.9   Projections on the Outside of the Product

Up to now, we have dealt only with projections that can be withdrawn easily, as the mold opens; the product is ejected simply by pushing it off the core or out of the cavity. Many projections, however, cannot be ejected that simply, because they enter "sideways" (i.e., at right angles or at any angle to the direction of ejection) into the product and would resist the simple pushing off (or out). There

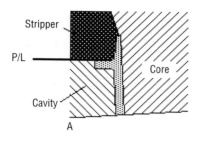 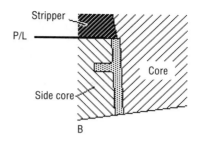

Figure 4.50 Two cross sections of a pail rim: A. projection is within the stripper ring, and B. projection is formed by side cores (Note that mold features such as cooling, venting, etc., are not shown)

is nothing wrong with such designs, but the product designer must understand that these projections may present some complications in moldmaking, increase mold costs, and often increase the cycle time. There is a large number of possibilities; the following sections will discuss a few common examples.

Figure 4.50 shows a portion of the same product (e.g., the rim of a pail). In Fig. 4.50A, the wall ends at the P/L, and the projection is within the stripper (ring). Because of the draft of the projection within the stripper, the product can be simply ejected from the core, since the cavity withdraws (opens) without "catching" the product. (NOTE: Air pressure may be required to lift the product out of the stripper if the extension is long and could hold the product in the stripper.)

In Fig. 4.50B, the product does not allow a taper in the stripper as shown in Fig. 4.50A; the wall ends at the P/L at a location different from example A. The projection in B is now in the location where the wall was in A, and must be formed by side cores. This is a considerably more complicated and expensive method of building a mold.

The point made here is that a small change in the product design—the simple addition of a draft angle—will permit the product to be made in a much simpler and less expensive mold. There are many other occasions where simplification similar to that suggested above can be applied.

4.4.4.9.1 Outside Threads

Outside threads are outside projections, and will always need at least two side cores to free the molded piece. In some cases, the mold designer may want to use more than two side cores (three or four) to reduce the stroke of the side cores. In the thread section of the product, there are always as many witness lines visible as there are side cores.

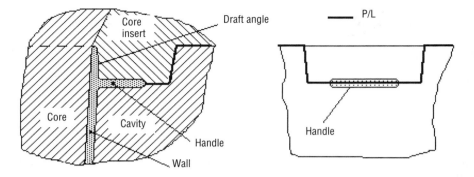

Figure 4.51 Cross section (left) of a container wall with a handle that is designed along an offset parting line (end view, right)

4.4.4.9.2 Other Outside Projections Products in this group include handles, etc., which would hinder the straight withdrawal of the product from the mold cavity. In some cases, side cores can be used, rather than split cavities. When reconsidering the shape of the product, the designer may find ways to design the projection along the P/L, even if it requires an offset P/L dropping to the level of the projection (Fig. 4.51). Such offset P/L costs more than a straight P/L but still can be much less expensive than either side cores or split cavities.

As in Fig. 4.50, there must be a draft angle to allow the product to pull straight out of the core. By selecting this design and agreeing to the concession of a draft angle, the use of a very costly side core to create the handle has been avoided.

Note that there will be a witness line where the insert meets the vertical wall. However, there would also be witness lines if the handle were produced with a side core or with a split cavity.

When the designer is in doubt and needs to visualize the area in question, a simple (and very low cost) method is to model this area with modelling clay (or "Plasticine"). The more sophisticated designer can use computer modelling, if accessible.

4.4.4.10   Undercuts on the Inside of the Product

Undercuts on the inside of a product are usually:

- Inside threads,
- snap-on ribs (for lids of containers), or
- major undercuts that are part of the product shape.

4.4.4.11 Inside Threads

Inside threads usually occur in closures but also may be designed in many technical products such as plastic pipe hardware, etc. The planned method of manufacture (molding, or machining after molding) should be considered. The mold could be much simpler (and considerably less costly) if the product is threaded after molding; the designer should be aware of the production quantities required. As we stated before, it is more important to arrive at the lowest total cost product than at the most elegant method of manufacture. If the thread *must* be molded, we again have two choices of how to remove the product from the core: by unscrewing or by stripping.

4.4.4.11.1 Unscrewing Threads are usually designed to standards and, as a rule, consist of more than one pitch (length). In many screw caps for bottles, jars, tooth paste tubes, and technical closures, two to six pitches are quite common. Within reasonable limits, there is usually no problem with the number of turns that are required to unscrew the cap, except that, the more turns, the larger will be the actuating mechanisms required in conventional unscrewing molds. Also, more turns mean longer products, more plastic, more molding, and more unscrewing time (longer cycles). In many closures, it has been shown that a thread length of one or two pitches is sufficient for good holding power and tightness of the closure.

This is an area the designer must carefully consider because the wrong decision can become very costly in the long run. There are several unscrewing methods; they will not be described here. However, all depend on features that are not part of the product design but which must be indicated on the product drawing. This refers to the method by which the product is held in the mold (or the unscrewing mechanism) after the mold has opened to remove the product from the core which has created the screw thread.

The most frequently used ejection methods are either rotating the core or the stripper ring; the product must be held so that it can be unscrewed while the product turns relative to the core, and retracts. This usually entails *teeth* or *ratchets* on the underside of the wall of the screw cap. In one system, the product requires *ribs* or other projections on the outside, where an external unscrewing device can engage to grab the closures to remove them from the cores. The design of these aids for unscrewing should be discussed with the *mold* designer for this project.

From the foregoing discussion, it becomes clear that any unscrewing method requires either complicated molds or special machines. The molding cycles are also slower than comparable products that need not be unscrewed. In low

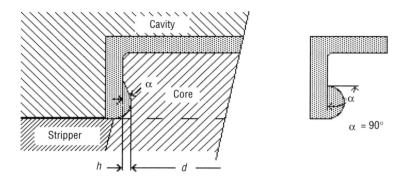

Figure 4.52 Cross section of a product ejected using a stripper to push the plastic over the "hump" in the core; the greater angle a, the greater the difficulty in stripping

quantity applications, the core could be also removable from the mold and unscrewed from the product outside of the mold, by hand or by using fixtures.

4.4.4.11.2 Stripping **The designer must consider whether the product can be stripped from the threads.** Stripping is the easiest (and often the lowest cost) solution for ejecting molded products; however, the ease of stripping depends on many equally important factors.

The theory of stripping is quite simple. As the mold opens, and after the cavity has moved away from the core side, the ejection starts, caused by the stripper moving forward. In doing so, the plastic product is pushed over the hump in the core (Fig.4.52); this causes the plastic to expand so that the portion that is inside the groove in the core can slip out of the groove.

4.4.4.11.2.1 Influences Affecting the Ability to Strip a Product The greater the angle a, the more difficult is the stripping action. This is quite obvious: a sharp angle (a = 90%, see Fig. 4.52, right) makes ejection virtually impossible. The projection would act like a hook, and the plastic would shear off in the groove rather than pull out. Screw threads could have a 90% angle if absolutely required, but that would definitely call for unscrewing; stripping would not be possible.

The greater the height h, the more difficult is the stripping action. This, too, seems obvious, since the greater h becomes, the greater a becomes. However, there is an important relationship between the height h, the diameter d, and the ease of ejection. As the plastic is dragged out of the groove, it stretches and is stressed in tension. The amount of stretching can be calculated by using the circumference C of the circle, with the diameter d, $(C = d\pi)$. The percentage of stretching f is directly proportional to the height h and inversely proportional to the diameter d; or f is proportional to h/d.

In other words, the greater the diameter d, the easier the product will slip out of its groove in the core. A product made from a relatively stiff plastic could be stripped if d is large, but it would break during ejection if d is small. The type of plastic, and its modulus of elasticity E and tensile strength are, therefore, important factors in determining the sizes and shapes of an undercut planned to function in some manner, such as to hold (snap onto) some other product or to form a screw thread.

Note that many screw threads, and especially threads for closures, have standardized shapes, which originally evolved from threads on glass bottles and jars. Such standards should be followed, and their large tolerances used to obtain maximum benefits for plastic threads. Also, note that the finish of the groove also has some influence on the ease of stripping, especially with materials requiring high polish for good ejection.

4.4.4.11.2.2 Number of Threads (Pitches)   The easiest and best way to strip threads is to specify only one thread (one pitch) so that the stripped thread will not slide into an adjoining groove but continue to slide uninterrupted off the core. By considering the factors influencing stripping as described above, the designer will succeed in creating a screw cap with the proper shape of the cross section of the thread (projection) and a suitable h/d ratio so that the product made from even a stiff plastic can be readily stripped, without the need for an unscrewing mold.

4.4.4.12   Other Projections

Almost anything can be molded in plastic; it is just a question of how much a mold will cost and how efficiently it will run. It is virtually impossible to describe all of the different shapes that can be molded, and also all of the problems that can be encountered. A few more common designs are shown in Fig. 4.53.

Figure 4.53A shows a typical, frequently found design (in many variations) which requires a substantial hook below the top surface of a product. Figure 4.53B illustrates a practical method to produce this shape by molding; however, it is quite expensive to achieve, since it requires more mechanism and permits less cooling of the core. During ejection, the slide ejector advances while it also moves to the left, thereby clearing the hook before the product is finally ejected.

Figure 4.53C shows how the product design can be altered to greatly simplify the mold. However, this redesign requires a change to the product: a

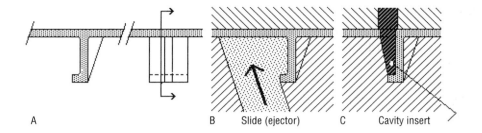

Figure 4.53 A product (A) requires a hook below the top surface which can be produced using a slide ejector (B) or a cavity insert (C)

small, usually rectangular opening must be left in the top surface of the product to permit the cavity insert to pass through and form the inside of the hook. This method requires just an "ordinary, up and down" mold, without moving ejectors, etc. If such an opening is not objectionable for the appearance or function of the product, this method is the preferred design because it is much less expensive than any alternative.

4.4.5   Other "Difficult" Shapes

The product designer should understand that there are many different methods used to free seemingly trapped undercuts so that the product can be ejected. The following two examples illustrate what is involved to mold such shapes, and that such solutions are possible but more expensive than designs without such features.

4.4.5.1   Freeing Undercuts in an Overcap

The central, circular projection in Fig. 4.54 ends in an enlarged bead, which is trapped by the surrounding mold steel. It would be impossible to eject the product using conventional methods. By using a two-stage ejection method, the bead can be freed before ejection is completed.

As shown in Fig. 4.54 (right), the sleeve and the stripper first move together (1+2) until the bead is above the top of the core. The stripper then continues alone (1), and strips the product off the core. The bead is free to deform (elastically) into the space previously occupied by the core, as indicated by the smaller arrows. On completion of ejection, the bead springs back to its molded shape.

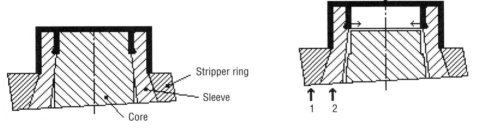

Figure 4.54 Cross section (left) shows a product with a circular projection requiring a two-stage ejection method

In this design, as with the shape of screw threads, etc., the shape of the bead must be shaped so that it will slide out of its groove. Also, the relation between the height of the groove h, the diameter d, and the molding material is as important as that shown for stripping screw threads.

4.4.5.2   Collapsible Cores

In principle, collapsible cores are the same as the slide ejector method shown in Fig.4.53, but they are used not just for one specific area, as illustrated, but for larger undercuts, which may be along much of the inside of the product. The mold becomes very complicated in design, expensive to build, and difficult to maintain; also, it is usually difficult to provide good cooling for the moving mold parts. All these factors result in a higher product cost.

References

1.   Rees, Herbert. *Mold Engineering*. Munich: Hanser Publishers, 1995.
2.   Ogorkiewitz, R.M. *Engineering Properties of Thermoplastics*. New York: Wiley-Interscience, 1970.
3.   *Modern Plastics Encyclopedia*. Heightstown, NJ: McGraw-Hill.

5   Designing for Assemblies

5.1   Mismatch

Matching of the outside of two molded pieces, or of one molded piece with a second piece made from another material, is virtually impossible, in view of the positional and diametrical manufacturing tolerances. The total variation of dimensions (and the mismatch) can be as high as the sum of the tolerances of the number of pieces involved. Matching is particularly difficult when pieces come from more than one cavity.

It is good practice to prevent unsightly mismatch between top and bottom parts by deliberately making one of the two parts larger along the matching face so that, even in case of mismatch, a more pleasant appearance can be achieved. The illustrations (Fig. 5.1, exaggerated) show one method for avoiding poor appearance due to mismatch caused by manufacturing and shrinkage variations.

Only one design style of meeting surfaces is shown, but the principle applies to any design. (In other illustrations in this chapter, the matching of assembled

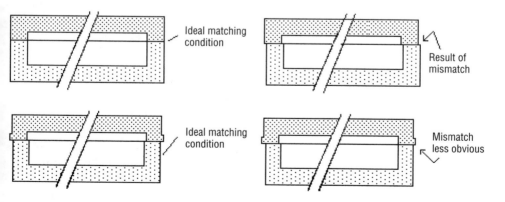

Figure 5.1 Poor appearance as a result of mismatch can be avoided by deliberately designing one part larger than the other

pieces will not show such provisions to hide mismatch but will assume that ideal match has been achieved.)

The following sections cover some of the most common assembly methods for plastic pieces and their effects on product design.

5.2   Screw Assembly

This description applies equally to assemblies using threads tapped in the plastic or to self-tapping screws. Screws are generally used for larger assemblies, such as enclosures for electrical apparatus, etc., and for any size assembly that must be taken apart for servicing or into which are fastened other components which may have to be removed for servicing, etc.

The number of screws depends on the design (the shape) of the pieces to be joined. Screws must ensure that the parts are held together under operating conditions of the assembly. If there is a seal placed between the two assembled parts, the distance of the screws must be such that the deflection of the sealing surfaces, caused by compressing a (soft) seal (rubber, cork, O-rings, etc.), will not be so great that the surfaces can separate and cause leaks, or let air (dust) enter.

5.2.1   Holding Power of Screws

In mechanical assemblies of any kind, the screw force depends on the amount the screw is preloaded (stretched) when torquing while compressing the assembly. Sufficient preload also ensures that the screw will not loosen as long as the preload prevails. With plastics, this reliance on preload is not possible, for two reasons:

1. Plastics in general are not strong enough to allow a screw to be torqued (and stressed) as high as when used to hold stronger materials such as steel. When torquing too much, the female thread in the plastic will strip, and the result will be loss of holding power. In plastic assemblies, therefore, screws can only be tightened such that the plastic thread will not deform (by preloading) beyond its permissible limits (in shear and/or deflection).
2. As shown earlier, most plastics will *creep* to a lesser or greater amount if a load is maintained for a considerable length of time. Gradually, the original compressed length will conform to the less

stretched, or unstretched, length of the screw and will lose some or all of its original holding force. The preload in the plastic thread itself will gradually disappear, and the screw will be held against unscrewing only by the friction of the screw in the plastic. (This friction power is higher for self-tapping screws than for tapped holes in the plastic.) It is, therefore, important to determine the creep values for the selected plastic, at the temperature the product will be used, before selecting a fastening method.

Relying on friction to hold the screws in place can often be satisfactory when assembling plastic components to each other; this is the case with most enclosures. However, where providing good preload is important, steel nuts can be inserted in the mold or at assembly. Even when steel nuts are used, the designer must remember that the plastic between the screw head and the nut is subjected to the load, which causes the plastic to creep and will lessen, or even eliminate, the tension (preload) in the screw.

Where continuous preload is required, as in the case of electrical contacts, metal (steel) nuts inserted in the plastic product will ensure that the screw can be tightened only against metal, maintaining the pressure on the electrical components. Poor tightening of an electrical contact, particularly if the currents passing through it are high, will heat up the joint and may soften the surrounding plastic, or even cause a fire.

The proper design is to allow such inserts to project by an amount h ($h \geq 1$ mm) above the plastic surface so that the screw will only exert force on the insert, not on the plastic. In Fig. 5.2, the screw can be properly tightened and will maintain its holding power.

The projection h has another advantage: it assists in holding the insert in the mold part where it is inserted before the mold closes. Good fit on its circumfer-

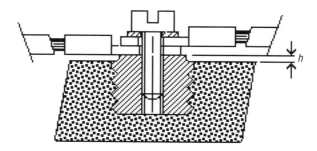

Figure 5.2 A steel nut inserted in the plastic part projects a distance h above the surface and will allow the screw to contact only metal, ensuring its holding power

ence will prevent the plastic from flashing over the metal, which would then require cleaning after molding.

5.2.2   Number of Screws

The number of screws should be held to a minimum. With increasing number of screws:

- Mold cost increases,
- cost of the molded product increases,
- product assembly requires increased labor,
- screw cost increases, and
- cost of required nuts or inserts increases.

However, with decreasing number of screws:

- Holding power may be insufficient, and
- a loss of parallelism and, therefore, sealing capability may result.

The designer must get an indication of, or estimate, the forces that will affect the joint, and also any internal pressures within the assembly which may tend to separate the joint and cause it to leak or allow air (and dirt) to enter. Once such forces are known, the designer can calculate the holding forces per screw (and determine their size) and design the product so that the plastic between the screws is strong enough to withstand these bending forces without creating a gap in the seating (or sealing) surfaces.

5.2.3   Size of Screws

From the forces estimated as described above, the designer will have some indication of the minimum size screw required to hold the assembly together. However, there are some practical limitations:

- If the screws are very small (1.5-, 2.0-, or 2.5-mm diameter, or in the USA #4, #5, or #6), they are more difficult to handle, both in assembly and in maintenance, than larger screws.
- The holes into which these small screws will enter must be even smaller. (Hole diameters and shapes are specified by the screw supplier and are also shown on screw charts.) Note that there is a difference in hole size and shape 1) for tapped threads in the plastic

and 2) for self-tapping screws. There are even different holes required for different makes of self-tapping screws. These holes are produced using very small and fragile core pins; they can be a constant source of trouble in the molding operation and, consequently, will eventually increase the product cost.

Although it cannot be said that such small screws should *not* be used, a good designer should be aware that, even if such a small screw would do the required job, he or she should, for practical reasons and provided the space is available, select larger screws of at least a 3-mm diameter (#8).

· The cost of a screw is not necessarily based on its size but on its commercial demand. A rarely used, small screw may cost more than a frequently used, larger one. The designer should look into the cost of the screw and determine whether the required size is available at all.

Standard screw charts from suppliers show all available standard sizes (diameter and length). Special sizes can be very expensive unless the quantities required are large enough for the supplier to custom make them at a reasonable cost.

· The designer should *never* specify that a standard screw be *modified* for use with an assembly. This could be very expensive, especially in a mass-produced item.

Screw size considerations will affect the design, since they affect the length of the hole through which the screw must pass. The *screw length* (under its head) is the length of the clearance hole plus the amount the screw enters the (tapped) hole.

The length of effective engagement of the screw thread is that length which actively and effectively engages the plastic. In most screws, particularly in self-tapping screws, there could be a lengthy point (lead-in taper) which does little for holding but which must be considered. This point can be of different lengths and shapes for various self-tapping screw designs. The designer can either specify the type of screw required for the job or make the hole for the worst case.

The length of effective engagement for steel is about equal to the screw diameter. In plastic, it should be greater—about two to three times the screw diameter. The designer should follow the recommendations provided by the screw manufacturer and the plastics supplier for the proper, effective engagement length for the type of plastic and screw planned.

5.2.4   Screw Holes in Plastic

For reasons of appearance, and also because of the available screw length, the screw head is often located below the surface of the plastic. Therefore, the hole in the plastic must be "counterbored" and will be specified as such, even though it is molded, not counterbored by machining.

There are many considerations in specifying the shape of the counterbore. In molding, every hole in a plastic piece is created by a pin (the core pin). The core pins are often subjected to unbalanced side forces, caused by high injection pressures and poor flow conditions in the mold as a result of the gate location, which will deflect these pins.

The length of the hole will also depend on the available standard screw length. The following examples present a few typical cases.

Example 1 A relatively shallow top ("top" in this context means that part where the screw head is located).

Section at screw head

LS = Standard screw length
LC = Length of clearance holes
LB = Length of counterbore
LT = Length of screw hole
LE = Length of engagement

Example 1 illustrates several points to consider:

LS Standard screw length LS should be a *standard* screw length, regardless of whether the screw requires a tapped hole or is a self-tapping screw.

LE Length of effective engagement   Screws in plastic should have an LE 2 to 3 times the screw diameter d.

LB Length (depth) of the counterbore LB is dependent on the overall design parameters. Where appearance permits, there is no need for a counterbore, and the screw head will seat on the flat top surface. In many designs, however, the screw head is desired below the top surface; LB is then usually greater than the height of the screw head.

LC Length of the clearance hole   The clearance hole is usually produced by the same pin that makes the LB. The longer the hole ($LB + LC$), the longer is the pin that is subjected to side forces, and thereby bend or "core shift" during injection, which may create a severe mismatch with the hole into which the screws must thread.

LT Length of the screw hole   There is no set rule for this length (nor for the *LB* or *LC*). It is usually determined from the design requirement and the availability of a suitable screw.

LT must be greater than *LE* plus any pointed portion of the screw; otherwise, the screw would bottom in the hole and 1) make it impossible to tighten the joint and 2) could split the plastic. If the hole will be tapped (rare today), *LT* must be deep enough to permit a run-out for the tap. For self-tapping screws, the difference between *LT* and *LE* depends on the selected screw. In either case, the designer must use the recommendations for proper hole dimensions from the screw manufacturer's charts.

It is practically impossible to provide a uniform thickness of plastic around the screw holes. The seats of the matching parts should have about the same shape. The lack of uniformity in the plastic thickness creates several bad effects, as was discussed in Section 4.4.4.8 on hubs, p. 76.

The hole diameters in the top part, to clear the screw head and the shank, should be calculated so that the screw can be assembled even when all tolerances of top and bottom go in the "wrong" directions and create a maximum mismatch between the two pieces.

Example 2   A relatively deep top: A. short screw with long counterbore, B. short screw in recess, C. long screw with long screw hole, and D. long screw in short holes.

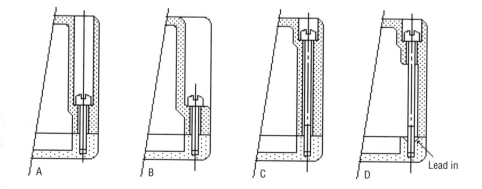

The sketches in example 2 illustrate screw assembly designs While the designs are possible or practical for the product in example 2, each highlights some of the problems the designer faces with the required design: a *deep* top (cover) screwed to a base.

In example 2A, a short, standard screw is easily obtained in the diameter required, regardless of whether the hole in the base is tapped or a self-tapping screw is used. The problems here are:

1. The exceptionally long counterbore, requiring a core pin which is easily deflected and difficult (or impossible) to cool internally, and
2. the positioning of a short screw in a long counterbore during assembly is undesirable, because it is difficult for a mechanic to hold the screw so that it enters centrally into the hole (except a steel screw, which can be held by a magnetized screw driver). Even so, this design is occasionally used.

The designer could ask the end user if it would be acceptable to have a recess in the side of the top, as shown in example 2B. In this case, the problem of cooling the long pin disappears; the recess is now part of the cavity block itself, and, in view of the large mass of the cavity block, there is no cooling problem. There is no assembly problem when positioning the screw into its hole from the outside. This is a preferred design. The question now is whether this design is acceptable for aesthetic reasons.

The design in example 2C has several flaws:

1. A long, thin screw of this size may not be standard and, therefore, becomes a "special" which should be used only as a last resort.
2. A core pin of the size required to create such a long hole in the plastic must be avoided in the mold under all circumstances. The pin is much too thin for its length and will deflect even more easily during injection than in example 1 above.
3. Such a core pin cannot be cooled using any conventional method. The molding cycle must be slowed down so much that the productivity of the mold becomes very low.

The fact that these flaws exist does not mean that design C has never been used, but it should be avoided because of the low productivity of the mold, the increased cost of the molded piece, the possible need for a special screw, and the high cost of upkeep of the mold. As a general rule, *a core pin should not project (be unsupported) for more than three to four times its diameter.*

The design in example 2D avoids the problem of the long hole (or core pin). The core pin is now short, and the cooling and deflection are less of a problem. However, it has its disadvantages:

1. A special, long screw may still be required, and
2. insertion of the screws during assembly is "blind," that is, in assembly it is difficult to find the hole where the screw has to enter for threading. A tapered lead-in as shown can be helpful.

This design is sometimes used but does not take into account the "groans" of the end user, who may be faced with reassembling the enclosure after servicing and then experience trouble when repositioning the screw.

Note that the above considerations apply not only to the few example shown but also to any product requiring long holes.

5.2.5   Long Holes in Plastics

In many designs (not only in the few cases described above), the length of any hole can be produced using two pins which meet approximately at the center of the hole. In such a case, the potential for deflection of the core pin is greatly reduced. Remember that the deflection at the end of a uniformly loaded beam is proportional to the *third* power of its free (unsupported) length. In other words, all other things equal, a pin of 20 mm in length will deflect eight times more ($2^3 = 8$) than a pin that is only 10 mm long ($1^3 = 1$).

Because of machining tolerances, it may be difficult to match the center lines of such pins—one of them located in the core, the other in the cavity.

Although this long hole is what the design in Fig. 5.3A requires, the mold (core) pin would be very long and easily bent.

The designer can create the long hole using two pins meeting halfway, as shown in Fig. 5.3B. This design will greatly improve the length/diameter ratio

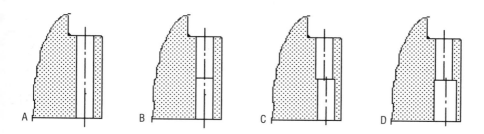

Figure 5.3 Examples of various core pin designs: A. one long core pin, B. two core pins meeting halfway, C. Two core pins that do not meet correctly, and D. Larger core pin on bottom

but will increase the mold cost by requiring two pins instead of one. However, this design does not account for diameter and positional tolerances of the pins.

Figure 5.3C illustrates what could actually happen if the two pins don't meet perfectly and, instead, create a restriction *and* a step, which may make insertion of the screw more difficult or even impossible.

Figure 5.3D shows (exaggerated) how the lower pin is made larger so that, even in the worst case (when all tolerances go in the "wrong" direction), there will be no restriction or step on the side from which the screw enters.

Note also that the longer a small pin is away from its (cooled) support in the core or cavity, the poorer is its cooling (see also p. 78). Hence, a long core pin is not only weaker (more deflection) but also hotter, and will slow down the molding cycle. Larger diameter pins can be cooled internally using one of several methods common in mold design.

5.3   Other Assembly Methods

There are always new developments in assembly methods. Some of the more common methods are highlighted below. The designer should get the latest data from experts in the selected field.

5.3.1   Bonding

Depending on whether the plastic is to be bonded to another plastic, to metal, or to any other material, a different surface finish than that commonly used in molding may be required. A rough surface may be most desirable for an adhesive, but the designer must know that rough surfaces could create difficulty in ejecting the product from the mold, especially if the roughness is located on side walls or ribs with little draft angle. (Note that some plastics, such as polyethylene (PE), eject easier if the molding surface is not highly polished.)

The designer should contact an expert on plastics adhesives for help in determining the type of adhesive suitable for the plastics being considered. The expert also can describe the surface specifications suitable for bonding.

For some plastics, the use of certain solvents is sufficient for bonding two pieces. They dissolve the surface layer and will bond (under some pressure) if used before the solvent evaporates. Some solvents are dangerous to operators and the plant environment (unhealthy and/or explosive) and must be used with special care. Solvents and adhesives also may be prohibited for use if the product or assembly is used for food, drugs, or medical purposes.

5.3.2   Welding

5.3.2.1   Sonic Welding

The pieces to be joined are placed against each other for a short, controlled time, and under controlled force, between a "horn" and an anvil (Fig. 5.4). While the horn induces a high frequency vibration, the touching surfaces melt and fuse the assembly.

The melting process is facilitated by providing specially designed ridges, or "energy concentrators," to one of the matching plastic surfaces. The shape, size, location, and number of the energy concentrators is best specified by the welding equipment supplier.

The designer should use to advantage the expertise of the sonic welding equipment supplier. There are also books on sonic welding, available from the supplier, which show different methods of designing joints for sonic welding.

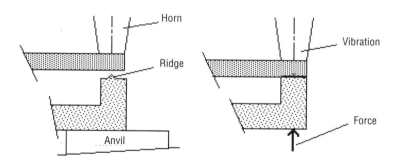

Figure 5.4 Pieces to be welded are placed between a vibrating horn and an anvil, and then forced together for a time until they are fused. A ridge is shown on the bottom part

5.3.2.2   Spin Welding

The spin welding method is applicable to circular surfaces which must be bonded. The two pieces to be joined are slightly pressed together while one of the pieces is rapidly rotated. The friction thus generated heats the plastic

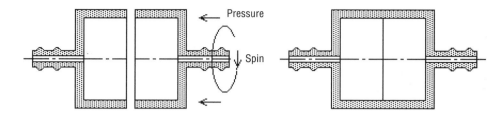

Figure 5.5 Two pieces of a nylon fuel filter are joined under pressure and spin welded

sufficiently to melt and fuse completely. A typical application is an automotive, in-line fuel filter made of nylon, schematically shown in Fig. 5.5.

The shape (cross section) of the surfaces to be joined may also have to be determined in cooperation with the manufacturer of spin welding equipment, and the plastic supplier.

5.3.2.3   Heat Welding

The plastic pieces are heated along the edges or surfaces to be joined until, on each part, a thin layer of the plastic melts enough so that, when pressed together, the two surfaces will fuse to make the end product. The method is rather crude and, dimensionally, is not very accurate, but it is used for joining large halves of industrial pallets, where fairly large tolerances are acceptable. Heat welding is also commonly used for joining plastic (PVC) rigid pipe.

Both materials to be joined should be of the same grade. The depth of melted surface may be anywhere between 1 and 3 mm.

5.3.2.4   Heat Sealing

Some bonding methods use applied heat, usually from microwaves. One example is to seal a plastic food container with plastic film.

The designer should approach an expert on heat sealing, and the supplier of the plastic contemplated for the job, to arrive at the proper design (area, finish, etc.) of the sealing surface.

5.3.4   Staking

For staking, one or a number of studs standing up on one part fit into matching hole(s) in the companion part. These studs are usually (but not necessarily) round and are somewhat longer than the holes into which they fit. The stud will then fill the hole and may leave no projection, or "dome," above the surface of the matching part. The studs should have a radius at the base (as shown in Fig. 5.6), but they could also be chamfered, provided the two corners of the chamfer are well rounded.

The volume of the stud, VS, is

$$VS = 0.25\ \pi \times DS^2 \times L \qquad (5.1)$$

and the volume of hole, VH, is

$$VH = 0.25\ \pi \times DH^2 \times H. \qquad (5.2)$$

VS must be larger than VH so that a dome can be formed. However, much of the holding force is generated by the cylindrical match of stud and hole.

The manufacturing tolerances of the sizes DS, L, DH, and H, and the tolerances governing their location in relation to each other, require that the stud be smaller than the hole. However, for proper staking, the *volume* of the stud, VS, must be at least as large as that of the hole, to fill it completely, and then some more, to create a dome over the stud and to apply sufficient pressure on the sides of the hole to provide the required holding force for the joint.

In another preferred design (Fig. 5.7), the holding capability is improved by giving the hole a reverse taper shape so that the stud will be wider at the staked end than at its base. This method eliminates the need for a dome.

When staking with a flat tool, the stud will end flush with the top surface. If the stud is originally too high (too much VS), the stud will compress more and/

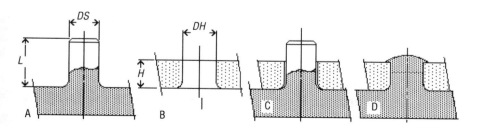

Figure 5.6 Typical staked joint: A. stud, with length L and diameter DS, B. hole, with length H and diameter DH ($DH > DS$), C. before staking, and D. after staking—the hole is completely filled by compressing the stud

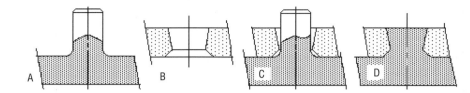

Figure 5.7 Staking by reverse taper: A. stud, B. hole with reverse taper shape, C. before staking, and D. after staking—the stud is wider at the base and has no dome

or the hole will be stretched by the force exerted by the staking tool. Plastic is compressible and can accept such forces as long as they are within the allowable stress limits. If the stresses are too large, the hole could burst.

Note that it is very important that every stud is well rounded at its base to reduce the effects of stress concentration and to prevent breakage of the stud. (See Section 5.4, Chamfers and Radii.) The hole in Fig. 5.7B must be made by two core pins meeting at the narrow section. The same rules apply as on p. 94, core pins meeting core pins.

Using a specially shaped staking tool, some plastics can be staked cold; others require elevated temperatures. With some brittle plastics, cold staking may not be possible, and the staking tool (and/or the plastic) must be heated.

Even though a staked joint can be used only once, and the pieces cannot be separated without destroying at least the part with the studs, the method is inexpensive and frequently used in mass production. Much work is often involved in developing the most suitable shapes for stud, hole, and staking tool before settling on a final design.

There is not much, if any, difference in assembly time when staking only one or (simultaneously) any number of studs per assembly; however, the number of studs (and matching holes) will affect the mold cost and may even affect the productivity of the mold. If ejectors are required under the studs, some of them may interfere with an efficient cooling layout for the mold, thereby causing the mold to run slower.

5.4   Chamfers and Radii—In the Plastic and in the Mold

The designer must be always aware that the mold presents the "negative" of the product shape. *This applies not only to core pins and holes but also to cavities and cores, in general.*

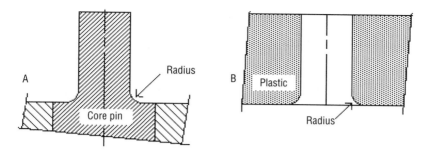

Figure 5.8 The core pin radius (A) is the "negative" of the radius in the molded part (B)

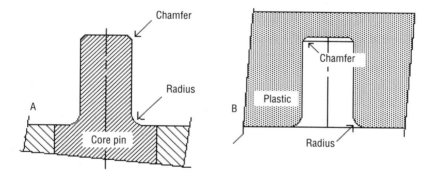

Figure 5.9 Chamfer on the end of the core pin (A) creates two stress risers in the plastic (or molded) part (B)

An inside radius in a steel part is easy to produce using any machining method (milling, turning, or grinding). This radius is also important for the life expectancy of the steel part, as it avoids stress concentration points (stress risers) and consequent breakage. The radius on the core pin will then produce a *radius* in the molded piece where, in normal design practice, a *chamfer* would be expected (Fig. 5.8).

On the other hand, a corner at the end of the core pin will be easier to chamfer than to make a radius; such chamfer will reproduce a chamfer on the inside of the plastic where normally a radius would be expected and desired. The (inside) chamfer in the plastic (Fig. 5.9B) now has two corners (stress risers); therefore, it is recommended that the designer specify the core pin chamfer to have well-rounded corners.

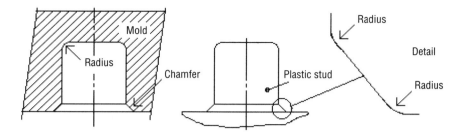

Figure 5.10 For plastic studs, the hole entrance should be well rounded, and the radius at the end of the hole should protect against stress in the cavity

However, the opposite is true when designing a plastic stud. The entrance of the hole (in the mold), which creates the plastic stud, is easy to make with a chamfer as shown in Fig. 5.10; but this point in the plastic should have well-rounded corners. It is far better to have the corner well rounded, as shown in Fig. 5.9. There should also be a radius at the deep end of the hole. This radius is a good protection against stress risers in the mold cavity.

If an ejector is used under the stud, which is desirable for studs longer than twice their diameter, it is difficult to make the end of the ejector come flush with the bottom of the hole. To ensure that the product will not hang up on the ejector pin embedded in the plastic, it is good practice to make the ejector pin slightly shorter, so that it will mold a projection on the plastic pin (Fig. 5.11). The allowable length L and tolerances of this projection must be clearly specified on the product drawing.

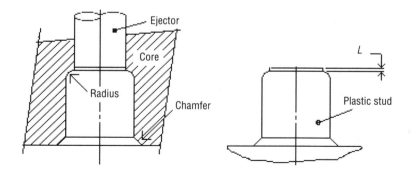

Figure 5.11 The ejector pin under the stud is short, causing a projection to be molded on the plastic pin of length L

As a rule, use only enough studs to make a satisfactorily functioning, safe assembly.

5.4.1   Design Assistance

For assistance with both cold and hot staking methods, the designer should contact a specialist (of the plastics manufacturer) and follow any suggestions regarding the length of stud projecting beyond the hole, the shape of the staking tools, and any necessary supports (anvils) under the studs.

6 Lettering and Other Distinctive Markings

Lettering and other markings are quite commonly shown on plastic products. Any or all of the following may appear:

- Manufacturer's name (and address),
- country of origin,
- recycling information,
- company logo,
- user instructions,
- patent and trademark information,
- cavity number (very important for quality control)
- date of manufacture, and
- batch number.

The designer must ascertain which of these data are required and indicate on the drawings where they are to appear. Some data may be legally required.

Some data could also be printed later, but at additional cost to the product. It is good practice to have as much data molded in as possible, provided it will not change in future batches. Printing is easier to change and may be the better choice if the same product may require different data in future.

Any molded-in data is either raised from or depressed in the product surface, so it must be located on surfaces where it will not prevent ejection from the mold. This is usually the top surface of the core or the inside bottom of the cavity.

If the information is shallow, 0.05 mm (0.001 in.), it may be produced using etching, EDM, or sand blasting, and even can be put on side walls with little draft. The product can then be ejected without problems. Such applications should be discussed with a molder or mold designer experienced in this field.

If more distinctive information must be shown on side walls, it can be produced using sliding cores, etc., but at a high cost, unless there are already slides required for other reasons, in which case such information could be added at little expense.

6.1   Shape of Engraving

Engraving is the most common method used to machine the information into the mold. As was mentioned earlier, the mold has the inverse (mirror) appearance of the product.

6.2   Raised Lettering

If the designer wants raised letters (or logos, etc.) in the plastic, the mold must be engraved; that is, metal (usually steel) must be removed where the information is to appear. This is the simplest and least expensive method, since only a minimum amount of metal is removed.

If the information is other than a standard style of lettering, art work must be supplied to the moldmaker so that the necessary masters for the engraving process can be produced. Art work is usually in the form of an enlargement of the required shape, such as a photo or a drawing. The ratio of enlargement depends on the job and can be anywhere from 2:1 to 20:1, as requested by the moldmaker.

6.3   Depressed Lettering

Although sometimes requested by a customer, the designer must be aware that depressed lettering is much more difficult to achieve (and costs much more), since now the metal must be removed *around* the information so that the lettering is left standing above the mold surface. Raised lettering is also far more easily damaged than engraved lettering.

There are other methods available for creating depressed lettering; for example, the lettering could be engraved into an electrode to be used for EDM. However, this too is more expensive than engraving directly into the mold steel.

One advantage of depressed lettering is that the lettering may be filled with paint after molding to add a color different from that of the plastic to make the information more visible. However, this adds even more cost to the finished product. Typically, typewriter keys can be made with two-color molding, with the lettering one color and the rest of the key another color. While this is done using any of a large number of molding methods, any one of them requires very expensive molds and machines and should be considered only when the product quantities are so large that this cost becomes a minor consideration.

As a rule, depressed lettering in a molded product should be discouraged. On the other hand, if other information must be added after molding, depressed lettering could be of advantage.

7 Checklist of Additional Product Requirements

The checklist for the designer, below, outlines additional product requirements not yet discussed which can influence a designer's decisions:

- What limits are there to overall product size, shape, and weight?
- Will the product be stacked for easier transportation and packing? How many units per stack?
- Does the product require stacking lugs for easy separation either in assembly or by the end user?
- Will the product be packed in standard quantities (100, 1000, dozen, etc.)?
- Are there size limitations for stack height or the size of cartons and boxes for packaging?
- Are standard box sizes to be used?
- Are there any other packaging requirements which can affect size and shape?
- Are there size limitations created by requirements for shipping in freight containers?
- How can the product be handled after molding and during secondary operation?
- Does the product need provisions to orient it for assembling, filling, or other secondary operations?
- Is the product sensitive to scratches or other damage in handling? (This could affect the method of manufacture.)
- Will there be a range of similar products (varying sizes, colors, lettering, etc.)?
- Is matching fit with other products required?
- Is nesting of the product required (one inside the other)?
- Is color match of several parts of an assembly important? (This may affect the selection of the molding method; for example, using a family mold which guarantees color match.)

8 Safety in Product Design

Safety is a very important part of the designer's job. When designing anything, the designer must be aware of all *foreseeable* possibilities that could cause harm and injury to the prospective user (and even the "misuser") and to others. The designer must also consider how the product will directly or indirectly affect the environment.

8.1 Foreseeable Areas of Risk

Foreseeable areas of risk, such as causing injuries or death, could be in the materials selection and in the shape or strength of the product. Consider the following examples:

- Sharp corners and projections,
- pinch points,
- possibility of being swallowed by infants and small children,
- harm caused by catastrophic failure of parts or all of the product,
- fire hazard, or
- failure to perform as advertised.

Because of some reluctance by the general public to the use of plastic instead of the conventional, older materials, the designer of plastic products must be especially careful. Consider the following typical example:

> Nobody has ever complained about the use of glass, which is known to shatter easily on impact, particularly when used for pressurized soft drink bottles. However, when the plastic industry proposed to make these bottles from PET, very stringent requirements were introduced. One such stipulation was that the filled and pressurized plastic bottle must withstand a drop of 6 feet (2 meters) without breaking. At this same time, the commonly used 2-liter glass bottles occasionally shattered when just tipped over on a flat surface.

Many accidents, particularly in North America, are exploited in the courts if they can be attributed to the failure of a product which caused bodily harm, or even only anguish. This "practice" is slowly spreading to other countries. The parties held responsible are often not only the seller of the so-called "unsafe" product; the designers who created it and the people involved in the making of such a product (the moldmaker and the molder) are often jointly held responsible and must defend themselves against such attacks as " . . . not to have *foreseen* the possibility of misuse of the product. . . ." In many cases, the designer or the manufacturer has been found guilty because of a " . . . *failure to warn* . . ." of possible risks when using the product, such as placing caution and warning information *on* the product, or in manuals supplied *with* the product. But, unfortunately, there are also justified claims if, for example, the designer did not care to select the proper material, or made some glaring design error, when calculating the strength of the product.

8.2   Responsibility and Liability

Before selection of a material, particularly if there are safety or health aspects in the application, the designer should demand to have all important specifications *certified* by the supplier, and if necessary, by an independent laboratory. This *may* help to protect the designer and the manufacturer of the new product against liability claims for nonperformance of the product or accidents resulting from the use of such material in a new application.

Of course, making and testing prototypes will go a long way toward proving the new design and its material selection. Unfortunately, these tests cannot prove beyond doubt how the product will behave under all kinds of adverse conditions.

Usually, the manufacturer wants to get the product out into the field as soon as possible, to take advantage of an early start in a competitive market. Unfortunately, this will often not allow much time for testing and proving the material to be used is safe, and thus brings a certain amount of risk for the manufacturer and the designer.

It is the responsibility of both manufacturer and designer to watch over the performance of the product and to follow up any bad reports. If necessary, the manufacturer must advertise the defect or notify known users and, if absolutely necessary, recall the products immediately upon becoming aware of problems, to avoid possible damage suits involving lengthy, costly litigation.

Index